THE
LINEAR AND DIGITAL
INTEGRATED CIRCUITS DESIGN
PRIMER

THE
LINEAR AND DIGITAL
INTEGRATED CIRCUITS DESIGN
PRIMER

A. SUDHAKAR

CHARLES RIVER MEDIA, INC.
Hingham, Massachusetts

A. Sudhakar. *The Linear and Digital Integrated Circuits Design Primer.*
ISBN: 1-58450-218-5

Acquisitions Editor: Brian J. Sawyer
Production: Laxmi Publications, LTD.
Cover Design: The Printed Image

CHARLES RIVER MEDIA, INC.
20 Downer Avenue, Suite 3
Hingham, Massachusetts 02043
781-740-0400
781-740-8816 (FAX)
info@charlesriver.com
www.charlesriver.com

This book is printed on acid-free paper.

Original Copyright 2002 by Laxmi Publications, LTD.
A. Sudhakar. *Comprehensive Linear and Digital Integrated Circuits Design.*
ISBN: 81-7008-212-9

Printed in the United States of America
02 7 6 5 4 3 2 First Edition

PREFACE

This book on linear and digital integrated circuits design covers all the topics of analog integrated circuits and digital integrated topics. For your convenience, the book has been divided into eleven chapters. The first seven chapters describe analog ICs and the last four chapters describe digital circuits.

Chapter 1 introduces the user to ICs and chapter 2 explains the fabrication details for different types of ICs.

Chapter 3 explains ideal op-amp characteristics and AC and DC characteristics of op-amp. Chapter 4 explains the applications of op-amp.

Chapter 5 explains phase-located loops. Chapter 6 explains 555 Timer IC. Chapter 7 explains the principles involved in analog to digital convertors and digital to analog convertors.

Chapter 8 explains the fundamentals of a digital system, starting from scratch. Chapter 9 explains some of the digital circuits, such as, Adder, ROM, EPROM, etc. Chapter 10 explains other important digital circuits.

Chapter 11 explains the logic families, which form the fundamentals of logic gates.

June 2002   **A. Sudhakar**

CONTENTS

1

Introduction to Integrated Circuits

1.1. BASIC DEFINITION OF IC

The integrated circuit, or IC, is a miniature, low-cost electronic circuit consisting of active and passive components that are irreparably joined together on a single crystal chip of silicon.

Advantages of IC

1. Miniaturization leads to increased equipment density.

2. System reliability is increased because soldering is avoided.

3. The cost is low because ICs are manufactured in a batch process.

4. Operating speed is increased.

5. Power consumption is decreased.

6. There is improved function performance.

1.2. CLASSIFICATION OF IC

ICs can be classified into different types depending on linearity, manufacturing process, or type of transistor used.

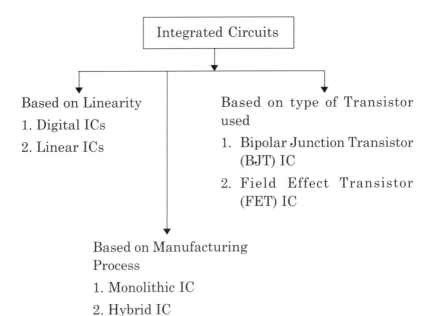

Based on Linearity

ICs can be classified as either digital or linear. Digital ICs have only two possible output levels. Logic gates and flip flops are examples of these. Linear ICs have any output voltage and will follow 'linearity' principle inside a specified range. Examples are operational amplifiers, timers, phase-locked loops, analog to digital converters (ADC), and digital to analog Convertors (DAC).

Based on Manufacturing Process

ICs can be classified as either monolithic or hybrid. Hybrid technology yields hybrid ICs. All others are monolithic.

Based on Type of Transistor Used

BJT ICs use Bipolar Junction Transistors. FET ICs use Field Effect Transistors.

2

IC Fabrication

2.1. PLANAR PROCESS

The silicon planar process of fabrication includes the following steps:

(*a*) Silicon Wafer (Substrate) Preparation

(*b*) Epitaxial Growth

(*c*) Oxidation

(*d*) Photolithography

(*e*) Diffusion

(*f*) Ion Implantation

(*g*) Isolation

(*h*) Metallization

(*i*) Packaging

The individual steps are explained in brief.

(*a*) **Silicon Wafer Preparation.** The basic material required for making the substrate, *i.e.* silicon, is cut into thin sheets, or wafers. This step includes the substeps like crystal growth, doping, slicing into wafers, and polishing and cleaning the wafer.

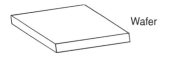

Fig. 2.1. Silicon wafer preparation.

(*b*) **Epitaxial Growth.** 'Epi' in Greek means 'upon' and 'teinon' in Greek means 'arranged.'

A small layer of silicon is grown over the substrate.

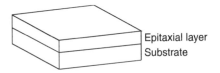

Fig. 2.2. Epitaxial growth.

Note. The thickness and other dimensions of the IC will be in mils.

where
$$1 \text{ mil} = \frac{1}{1000} \text{ inch}$$

$$\simeq \frac{2.5}{1000} \text{ cm}$$

$$\simeq 25 \ \mu\text{m}.$$

(*c*) **Oxidation.** Here an oxide layer is grown over the epitaxial layer. The SiO_2 layer formed by oxidation prevents diffusion of almost all impurities.

After oxidations the structure has 3 layers, *i.e.* the substrate, the epitaxial layer, and the oxide layer.

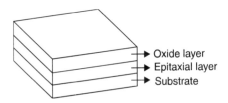

Fig. 2.3. Oxidation.

(*d*) **Photolithography.** The purpose of this step is to remove portions of the SiO_2 layer so diffusion can occur in selected areas.

The structure before applying the photolithography process is shown in Figure 2.4.

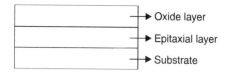

Fig. 2.4. Input structure for photolithography.

The photolithography process involves many steps which are described in detail below.

The first step is to make a photoresist layer above the oxide layer.

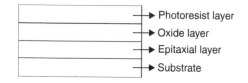

Fig. 2.5. Structure with photoresist layer.

The second step is to make masks, the locations of which are determined by the final structure. The masks will protect the photoresist layer from the ultraviolet light applied in step three.

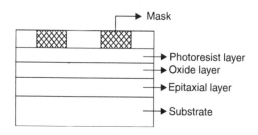

Fig. 2.6. Structure with masks.

Step three involves passing an ultraviolet light over the masks. Wherever the mask is not present, the photoresist layer is 'polymerized.'

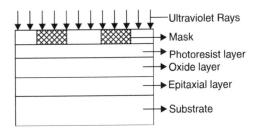

Fig. 2.7. Structure with mask layer
during ultraviolet exposure.

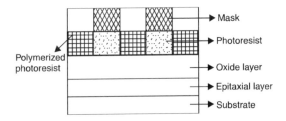

Fig. 2.8. Structure with mask layer
after ultra-violet exposure.

The structure is dipped into a solution of hydrofluoric acid in the fourth step. The masks and the non-polymerized portions of the photoresist layer will dissolve in the hydrofluoric acid solution.

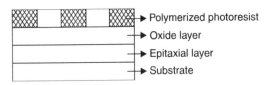

Fig. 2.9. Structure after dissolution in hydrofluoric acid.

The fifth step involves dipping the structure into a photosensitive emulsion. The oxide layer not protected by the polymerized photoresist will dissolve in the solution.

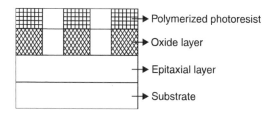

Fig. 2.10. Structure after dissolving in
photoemulsive solution.

The sixth and final step is to remove the polymerized
photoresist. At the end of this step, the oxide layer will be ex-
posed.

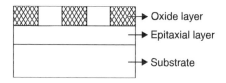

Fig. 2.11. Final structure after photolithography.

(*e*) **Diffusion.** Diffusion is the process by which the N-
type or P-type impurity silicon atoms can be diffused into the
epitaxial layer, through the holes in the oxide layer.

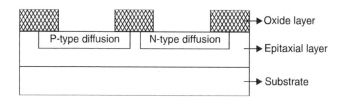

Fig. 2.12. Diffusion.

(*f*) **Ion Implantation.** In this process, an alternative to
diffusion, the epitaxial layer can be implanted with impurity
ions.

(*g*) **Isolation.** Since a number of different circuits are
manufactured in a single planar process, it becomes essential
to differentiate the circuits. This process checks whether any

short circuit is present between different circuits and, if so, the corresponding part is identified as unusable.

(*h*) **Metallization.** This process provides electrical metal contacts to the different diffused areas, where the terminals of the devices should be taken. Wherever the terminals should be short-circuited always, the metal contacts will be short-circuited and a single lead terminal will be taken 'out.'

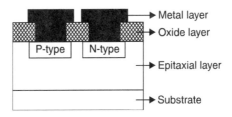

Fig. 2.13. Metallization.

(*i*) **Packaging.** The circuits manufactured in a single process will be scribed and cut down into separate structures. Each structure will be packed as a separate IC. The packaging will be used to give output leads to users.

There are three different package configurations available:

(1) To-5 Glass Metal Package

(2) Ceramic Flat Package

(3) Dual-In-Line Package (DIP).

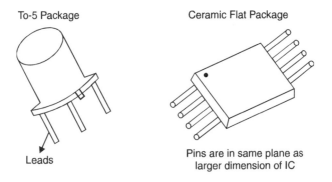

DIP Package

Pins are perpendicular
to larger dimensions of IC

Fig. 2.14. Different IC packages.

2.2. BIPOLAR JUNCTION TRANSISTOR FABRICATION

The first step is to make the substrate wafer. The next step is to grow the epitaxial layer. The third step is to oxidize the structure, thereby making the oxide layer.

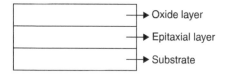

Fig. 2.15. Structure for BJT before photolithography.

The next step is photolithography during which portions of the oxide layer are removed.

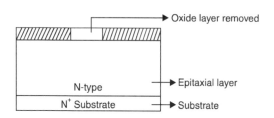

Fig. 2.16. Structure for BJT after photolithography.

Diffusion of P-type is done through the holes in the oxide layer.

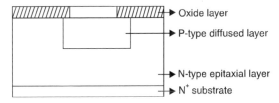

Fig. 2.17. Structure of BJT after base diffusion.

Oxidation is done once again to cover the P-type diffused layer.

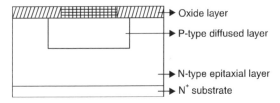

Fig. 2.18. Structure for BJT after oxidation for collector diffusion.

Photolithography is done once again.

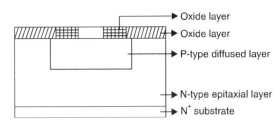

Fig. 2.19. Structure for BJT after photolithography for collector diffusion.

Diffusion of N-type is done.

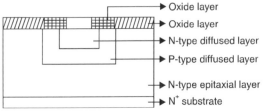

Fig. 2.20. Structure for BJT after collection diffusion.

Now we have a structure similar to an 'NPN' transistor. Three layers are formed by the N-type epitaxial layer, P-type diffused layer, and N-type diffused layer.

The next step is to isolate the different circuits manufactured in a single process. After isolation, metal contacts are introduced to these three layers through the oxide layer.

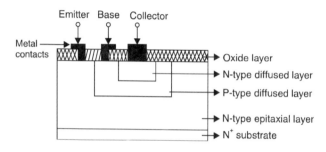

Fig. 2.21. Final structure of a BJT.

Then the circuit is packed in the IC package.

If the transistor to be manufactured is a 'PNP' type, then the type of impurity used in each level of diffusion will be of the opposite type, *i.e.* P-type epitaxial layer, N-type diffused layer first, and P-type diffused layer at the end.

2.3. FIELD EFFECT TRANSISTOR FABRICATION

Junction field effect transistor fabrication is similar to the fabrication of a BJT, with a few small differences. Since the basic steps have already been discussed, the final diagram of JFET IC is shown.

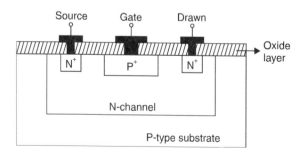

Fig. 2.22. N-channel JFET structure.

In the case of a metal oxide semiconductor FET (MOSFET), there should be a thin metal oxide layer between the gate and the channel. Since the basic steps of manufacturing are the same, the final structure is shown in Figure 2.23.

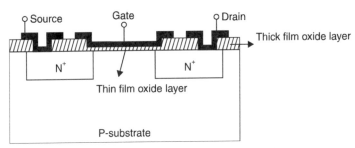

Fig. 2.23. Metal oxide semiconductor FET structure.

A MOSFET uses two different thicknesses for the oxide layer. Thick film oxide layer is from 5,000 to 10,000 Å and the thin film oxide layer is from 300 to 800 Å.

For the fabrication of opposite channel devices, the polarities of all blocks should be switched.

2.4. CMOS FABRICATION

CMOS (complementary metal oxide semiconductor field effect transistor) fabrication needs an N-channel MOSFET and P-channel MOSFET which are connected together as shown in Figure 2.24.

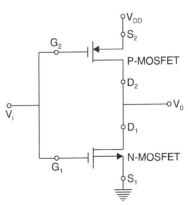

Fig. 2.24. N-channel and P-channel connection
for CMOS fabrication.

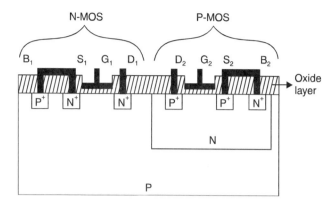

Fig. 2.25. Final structure of CMOS.

2.5. RESISTOR FABRICATION

Resistor fabrication is based on the fact that even semiconductors have some resistance. As the length of the semiconductor bar is varied, the resistance across the ends is going to be changed.

The simplest way to implement a resistor in IC form is to diffuse a P-type or N-type layer and take the resistance across the ends by placing metal contacts at the extreme ends of the layer.

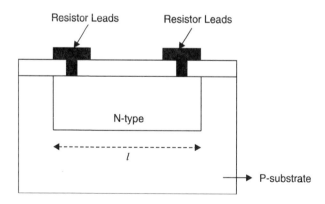

Fig. 2.26. Structure for resistor fabrication.

As l is varied, the resistance across the leads will vary.

2.6. CAPACITOR FABRICATION

Although a reverse-biased PN-junction diode can be used as a capacitor, IC capacitors are used less frequently because they are limited in range and performance.

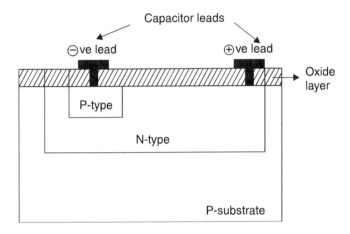

Fig. 2.27. Structure for capacitor fabrication.

2.7. MONOLITHIC DIODE FABRICATION

The two leads of a transistor will be used as diode leads.

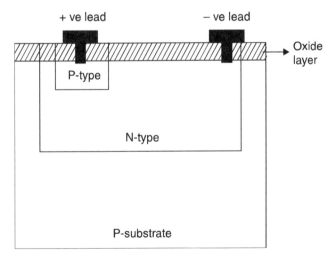

Fig. 2.28. Structure for diode fabrication.

Note. Monolithic is a Greek term.

'Mono' means 'single' and 'lithic' means 'stone.'

Meaning that this process is made up of a single series of manufacturing processes.

3

Operational Amplifiers

3.1. INTRODUCTION TO OPERATIONAL AMPLIFIERS

An operational amplifier, or op-amp, is basically a differential amplifier, *i.e.* it is used for amplifying the difference between two inputs. There are two signal input terminals and one signal output terminal. The input terminals are referred to as inverting input terminal (noted with a − symbol) and non-inverting input terminal (noted with a + symbol). (See Figure 3.1.)

The operational portion of the name comes from the fact that it can do any mathematical operation such as addition, subtraction, multiplication, division, logarithms, anti-logarithms, differentiation, and integration.

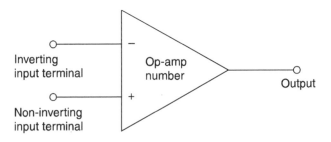

Fig. 3.1. Symbol of an operational amplifier.

The symbol of an op-amp can be drawn as a triangle with the inputs on one side and the output on the opposite side.

If the power supply is to be included, the symbol can be drawn as shown in Figure 3.2.

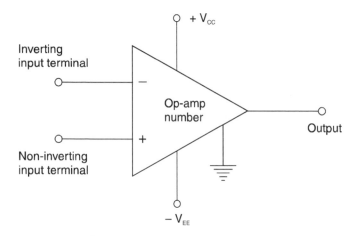

Fig. 3.2. Symbol of an operational amplifier
showing power supply connections.

With regard to power supply, the op-amp has three terminals: one positive $(+V_{CC})$, one negative $(-V_{EE})$, and one ground.

3.2. IDEAL OPERATIONAL AMPLIFIER CHARACTERISTICS

There are certain properties that an ideal op-amp should have. The ideal op-amp characteristics are described below.

(*a*) **Infinite input impedance.** If the input current is zero, then any signal source can drive this device.

(*b*) **Zero output impedance.** If the op-amp has zero output impedance, then the output voltage will be independent of the load resistance. Therefore, a single op-amp can drive many other devices.

(*c*) **Infinite open loop voltage gain.** Any minute input signal difference can be amplified to give a noticeable output.

(*d*) **Infinite bandwidth.** Bandwidth is defined as the range of frequencies within which the signal will be amplified to a sufficient level. Since bandwidth of an ideal op-amp is infinity, the ideal op-amp will amplify any signal from 0 Hz to ∞ Hz.

(e) **Perfect balance.** The output of an op-amp should be zero when the input voltages at both input terminals are equal to zero volts.

(f) **Infinite common mode rejection ratio (CMRR).** This causes amplification of the difference between the input terminals rather than the signals from the terminals themselves.

(g) **No output drift with temperature.** This causes the output to be independent of surrounding temperature and be dependent only on the input voltages.

(h) **Infinite slew rate.** Slew rate is a measure of time delay occurring between the output and the input of the op-amp. Slew rate is defined as the ratio of change in output voltage to change in time between the instants of the voltages measured. (See Figure 3.3.)

For example, if the square wave can be assumed as the input of the device, then the output should also be a square wave. But due to limitations of the device, the output takes some time to rise from one voltage level to another.

$$\text{Slew Rate } = \frac{\Delta V}{\Delta t} \tag{3.1}$$

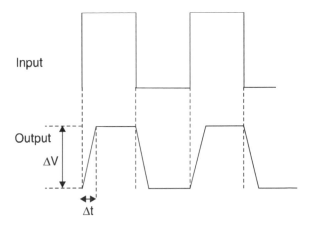

Fig. 3.3. Slew rate.

The higher the slew rate, the quicker the response from the device.

The ideal op-amp would have all of the above qualities and characteristics. In reality, none of these will be satisfied. The equivalent circuit can be drawn as shown in Figure 3.4 where A_v is the practical voltage gain of the op-amp.

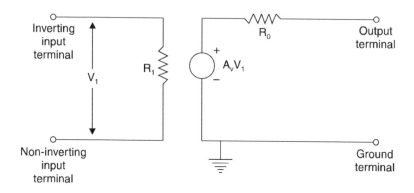

Fig. 3.4. Equivalent circuit of an op-amp.

3.3. DC CHARACTERISTICS

In this section we will discuss IC 741, which is a commonly used op-amp.

The pin diagram of IC 741 DIP package can be drawn as shown in Figure 3.5.

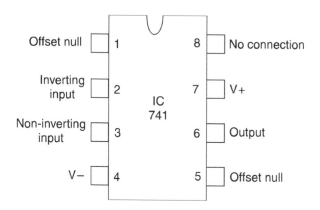

Fig. 3.5. Pin diagram of IC 741.

The important DC characteristics are the following:

(a) Input Bias Current (I_B)

(b) Input Offset Current (I_{OS})

(c) Input Offset Voltage (V_{OS})

(d) Thermal Drift

(e) Total Output Offset Voltage (V_{OOT})

(f) Differential Input Resistance (R_i)

(g) Input Capacitance (C_i)

(h) Input Voltage Range

(i) Common Mode Rejection Ratio (CMRR)

(j) Supply Voltage Rejection Ratio (SVRR)

(k) Output Voltage Swing

(l) Output Resistance (R_0)

(m) Output Short Circuit Current (I_{OSC})

(n) Supply Current (I_S)

(o) Power Consumption

(p) Gain-Bandwidth Product

(q) Slew Rate

(a) **Input Bias Current.** The input bias current is defined as the average of bias currents entering the op-amp through its input terminals.

Let the current entering the op-amp through the inverting input terminal be I_B^- and the current entering the op-amp through the non-inverting input terminal be I_B^+.

$$I_B = \frac{I_B^+ + I_B^-}{2} \qquad (3.2)$$

(b) **Input Offset Current.** The input offset current is defined as the algebraic difference between the bias currents entering the op-amp.

$$I_{OS} = |I_B^+ - I_B^-| \qquad (3.3)$$

(*c*) **Input Offset Voltage.** It is the amount of input voltage, that should be applied between two input terminals, to force the output voltage to become zero.

(*d*) **Thermal Drift.** All the DC parameters vary with change in temperature. This variance is referred to as thermal drift.

(*e*) **Total Output Offset Voltage.** This is defined as the output voltage taking into consideration input bias currents and input offset voltage.

To make this zero, the input offset voltage is applied. This principle is referred to as **offsetting, offset nulling, or DC offsetting**. For IC 741, pins 1 and 5 can be used for this purpose. By connecting a 10 kΩ potentiometer between pins 1 and 5, varying the potential, and applying this voltage along with $V-$ at pin 4, the output offset voltage can be made to equal zero.

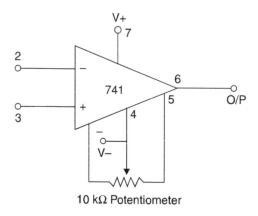

10 kΩ Potentiometer

Fig. 3.6. Circuit for offset nulling.

(*f*) **Differential Input Resistance.** This is also referred to as input resistance and is defined as the resistance between the two input terminals.

Input resistance is measured by taking the measurement of resistance from one input terminal with the other terminal connected to the ground.

(*g*) **Input Capacitance.** Input capacitance is defined as the capacitance between the two input terminals. It is measured by taking the capacitance at one input terminal with the other input terminal connected to the ground.

(*h*) **Input Voltage Range.** Input voltage range is defined as the maximum voltage that can be applied to both the input terminals in common.

(*i*) **Common Mode Rejection Ratio.** CMRR is defined as the ratio of voltage gain for differential input signal to the voltage gain for input signal present in common to both the input terminals.

$$\text{CMRR} = \frac{A_d}{A_{cm}} \qquad (3.4)$$

where A_d is the voltage gain for differential input signal

and A_{cm} is the voltage gain for input signal present in common to both the input terminals.

For an ideal op-amp, CMRR was defined to be infinity. If CMRR is high, it denotes that the op-amp is good at rejecting signals present in common to both the input terminals.

(*j*) **Supply Voltage Rejection Ratio.** SVRR is defined as the ratio of change in input offset voltage to the change in supply voltage, which caused the change in input offset voltage.

This parameter is also referred as Power Supply Rejection Ratio (PSRR) or Power Supply Sensitivity (PSS).

$$\text{SVRR} = \frac{\Delta V_{io}}{\Delta V} \qquad (3.5)$$

where ΔV_{io} is the change in input offset voltage

and ΔV is the change is supply voltage.

(*k*) **Output Voltage Swing.** This is defined as the variation in the voltage at the output terminal.

(*l*) **Output Resistance.** This is defined as the circuit resistance between the output terminal and the ground.

(*m*) **Output Short Circuit Current.** This is defined as the current flowing through the output of an op-amp when the output terminal is short-circuited to ground.

(n) **Supply Current.** The supply current is the current drawn from the power supply.

(o) **Power Consumption.** The quiescient power that must be consumed by the op-amp. for proper operation is the power consumption.

(p) **Gain Bandwidth Product.** The gain bandwidth product is defined as the product of gain of an op-amp and its bandwidth. It is equal to the bandwidth when the voltage gain is made one.

(q) **Slew Rate.** Slew rate is a measure of time delay occuring between the output and the input of op-amp. It has already been discussed in detail in section 3.2.

3.4. AC CHARACTERISTICS

The main feature in AC characteristics of an op-amp is frequency response. In the ideal op-amp, since bandwidth is infinity, the frequency response is constant at all frequencies, *i.e.* magnitude and phase of gain is constant for the ideal op-amp. But in reality, an op-amp exhibits some inconstant characteristics because of equivalent capacitance inside the op-amp.

A typical frequency response of an op-amp is drawn in Figure 3.7.

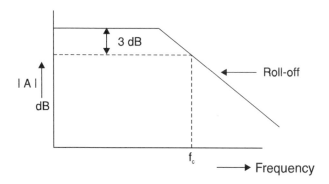

Fig. 3.7. Magnitude response of gain of an op-amp.

In some cases, there will be many stages of cut-off frequency and after each stage of cut-off frequency, the response gains a -20 dB/decade rate of roll-off.

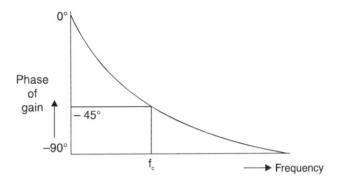

Fig. 3.8. Phase response of gain of an op-amp.

Compensation techniques can be used to increase the bandwidth.

3.5. INTERNAL CIRCUIT

A general op-amp has an internal circuit as shown in Figure 3.9.

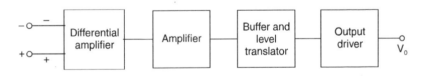

Fig. 3.9. Internal diagram of op-amp.

It consists of two stages of amplifier, which yield very high gain. The first stage is a differential amplifier, which produces an output equal to an amplified version of the difference between the two inputs.

The buffer and level translator stage usually consist of an emitter follower and a constant current source. The purpose of this stage is to save the previous two stages from overloading. The output driver stage results in a low output impedance.

The other stages present in op-amp will be for drift compensation and for frequency compensation.

3.6. FET Op-amps

A FET op-amp uses Field Effect Transistors in place of Bipolar Junction Transistors. FETs are easy to make in the form of integrated circuits and generally have high input impedance and low output impedance, which are desirable properties.

4

Operational Amplifier Applications

4.1. INVERTING AMPLIFIER

In an inverting amplifier, the input signal is given to the inverting input terminal and the non-inverting input terminal is connected to the ground. This configuration is shown in Figure 4.1.

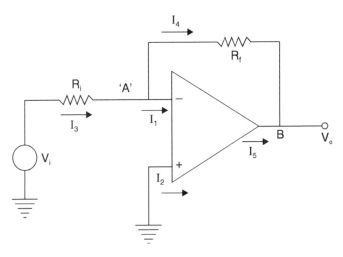

Fig. 4.1. Inverting amplifier.

The analysis will be done assuming ideal op-amp characteristics (refer to section 3.2).

The current from the op-amp input terminals is zero, *i.e.,* $I_1 = 0$ and $I_2 = 0$.

The gain of the op-amp is infinity. Therefore, to have a finite output, the input voltage difference between the two input terminals should be zero.

$$I_1 = I_2 = 0 \qquad (4.1)$$

The voltage at the non-inverting input terminal is zero volts because it is grounded. Therefore, the voltage at node 'A,' i.e., at the inverting input terminal, is also zero volts.

$$V_A = 0 \qquad (4.2)$$

Writing Kirchhoff's Current Law at node 'A' yields

$$I_1 + I_4 = I_3 .$$

Knowing that $I_1 = 0$ yields

$$I_4 = I_3 . \qquad (4.3)$$

Calculate the current through resistance R_i:

$$I_3 = \frac{V_i - V_A}{R_i}$$

Substitute equation (4.2) in this:

$$I_3 = \frac{V_i}{R_i} \qquad (4.4)$$

Calculate the current through resistance R_f:

$$I_4 = \frac{(V_A - V_o)}{R_f}$$

Substitute equation (4.2) in this:

$$I_4 = \frac{(-V_o)}{R_f}$$

Solve for V_o:

$$V_o = - R_f I_4$$

Substitute equation (4.3):

$$V_o = - R_f I_3$$

Substitute equation (4.4):

$$V_o = -R_f \left(\frac{V_i}{R_i} \right)$$

$$\therefore \qquad V_o = -\frac{R_f}{R_i} V_i \qquad (4.5)$$

This is the relationship between V_o, the output voltage, and V_i, the input voltage.

In equation (4.5), the minus sign indicates that there is a phase shift of 180° between input and output.

Equation (4.5) can be rearranged as

$$\frac{V_o}{V_i} = -\frac{R_f}{R_i} \qquad (4.6)$$

The conventional equation for voltage gain is

$$A_V = \frac{V_o}{V_i}$$

Substitute equation (4.6):

$$A_V = -\frac{R_f}{R_i} \qquad (4.7)$$

The magnitude of voltage gain is

$$|A_V| = \frac{R_f}{R_i} \qquad (4.8)$$

The input and output waveforms can be drawn as shown in Figure 4.2.

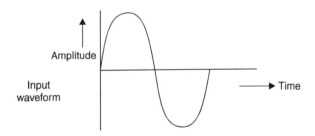

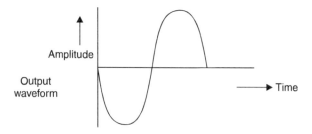

Fig. 4.2. Input-output relationship for inverting amplifier.

4.2. NON-INVERTING AMPLIFIER

In a non-inverting amplifier, the input signal is given to the non-inverting input terminal and the inverting input terminal is connected to the ground. This configuration is shown in Figure 4.3.

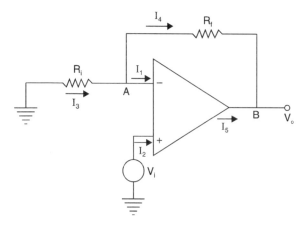

Fig. 4.3. Non-inverting amplifier.

The analysis will be done assuming ideal op-amp chara-cteristics, as done for the inverting amplifier.

The current from the op-amp input terminals is zero.

$$I_1 = 0 \qquad (4.9)$$

$$I_2 = 0 \qquad (4.10)$$

The gain of the op-amp is infinity. To have a finite output, the input voltage difference between the two input terminals should be zero. Since the voltage at the non-inverting terminal is V_i, the voltage at the inverting terminal must also be V_i to satisfy this property. Therefore, the voltage at node 'A' is

$$V_A = V_i. \qquad (4.11)$$

Writing Kirchhoff's Current Law at node 'A' yields

$$I_1 + I_4 = I_3.$$

Substitute equation (4.9) in this:

$$I_4 = I_3 \qquad (4.12)$$

Calculate the current through resistance R_i:

$$I_3 = \frac{(0 - V_A)}{R_i}$$

Solve for V_A:

$$V_A = -I_3 R_i \qquad (4.13)$$

Calculate the current through resistance R_f:

$$I_4 = \frac{(V_i - V_o)}{R_f}$$

Substitute equation (4.11) here:

$$I_4 = \frac{V_A - V_o}{R_f}$$

Substitute equation (4.12) here:

$$I_3 = \frac{V_A - V_o}{R_f}$$

Substitute equation (4.13) here:

$$I_3 = \frac{(-I_3 R_i) - V_o}{R_f}$$

$$\therefore \qquad I_3 R_f = -I_3 R_i - V_o \qquad (4.14)$$

From equation (4.13) in terms of V_A, I_3 can be written.

By doing so, equation (4.14) can be rewritten as:

$$\left(\frac{-V_A}{R_i}\right)R_f = (+V_A) - V_o$$

$$V_o = V_A + V_A\left(\frac{R_f}{R_i}\right)$$

$$= V_A\left[1 + \frac{R_f}{R_i}\right]$$

Substitute equation (4.11) here:

$$V_o = V_i\left[1 + \frac{R_f}{R_i}\right] \tag{4.15}$$

Equation (4.15) gives the relationship between the output voltage and the input voltage.

Unlike equation (4.5), equation (4.15) has no minus sign, which indicates that there is no phase shift between input and output.

Equation (4.15) can be rearranged as:

$$\frac{V_o}{V_i} = 1 + \frac{R_f}{R_i} \tag{4.16}$$

The conventional gain for voltage is

$$A_v = \frac{V_o}{V_i}$$

Substitute equation (4.16) into this equation to get the voltage gain:

$$A_v = 1 + \frac{R_f}{R_i} \tag{4.17}$$

The input and output waveforms can be drawn as shown in Figure 4.4.

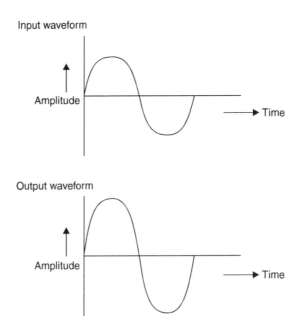

Fig. 4.4. Input-output relationship for
a non-inverting amplifier.

A special case of non-inverting amplifier called a **Voltage Follower** or **Source Follower** is a non-inverting amplifier where $R_f = 0$. Since $R_f = 0$, gain of this amplifier is equal to unity. This circuit is useful for impedance matching.

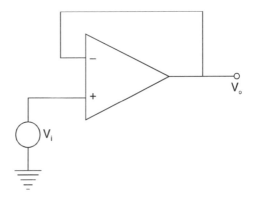

Fig. 4.5. Voltage follower.

4.3. SUMMER

In a summer, or summing amplifier, many inputs are given and the output is taken from the op-amp. If the inputs are given to the inverting input, then the configuration is an inverting summing amplifier. If the inputs are given to the non-inverting input, then the configuration is a non-inverting summing amplifier.

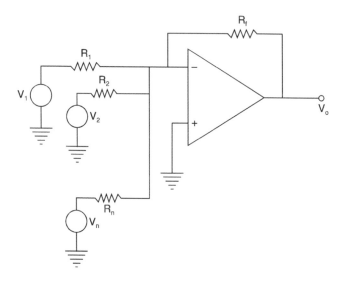

Fig. 4.6. Inverting Summer.

By applying the superposition theorem, it is easy to understand the operation of this circuit.

Assuming the effect due to input V_1, the output V_{o1} can be written as

$$V_{o1} = -\frac{R_f}{R_1}V_1 \qquad (4.18)$$

Similarly the output V_{o2}, due to input V_2, can be written as

$$V_{o2} = -\frac{R_f}{R_2}V_2 \qquad (4.19)$$

Similarly, the output V_{on}, due to input V_n, can be written as

$$V_{on} = -\frac{R_f}{R_n} V_n \qquad (4.20)$$

The total output V_0 of the system is the sum of the outputs due to individual inputs.

$$V_o = V_{o1} + V_{o2} + ... + V_{on}$$

Substitute from equations (4.18), (4.19) and (4.20):

$$V_o = \left(-\frac{R_f}{R_1} V_1 \right) + \left(-\frac{R_f}{R_2} V_2 \right) + ... + \left(-\frac{R_f}{R_n} V_n \right)$$

or

$$V_o = -R_f \left[\frac{V_1}{R_1} + \frac{V_2}{R_2} + ... + \frac{V_n}{R_n} \right] \qquad (4.21)$$

As a special case, if $R_1 = R_2 = ... = R_n = R$, then

$$V_o = -\frac{R_f}{R} [V_1 + V_2 + ... + V_n] \qquad (4.22)$$

which is the sum of all inputs, multiplied by $\frac{R_f}{R}$.

If $\frac{R_f}{R} = 1$, then

$$V_0 = -[V_1 + V_2 + ... + V_n] \qquad (4.23)$$

which is the sum of all input voltages.

If $\frac{R_f}{R} = \frac{1}{n}$, then

$$V_o = -\left[\frac{V_1 + V_2 + ... + V_n}{n} \right] \qquad (4.24)$$

where the output is the average of all input voltages.

4.4. DIFFERENTIAL AMPLIFIER

In the case of a differential amplifier, the input signals are applied to both the inverting input terminal and the non-inverting input terminal.

The configuration can be drawn as shown in Figure 4.7.

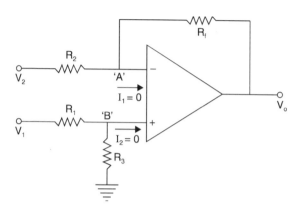

Fig. 4.7. Differential amplifier.

Here also, the ideal op-amp characteristics are assumed.
The voltage at node 'A', should be equal to voltage at node
'B.'

$$\therefore \qquad V_A = V_B \qquad\qquad (4.25)$$

Writing Kirchhoff's Current Law at node 'A' yields

$$\left(\frac{V_2 - V_A}{R_2}\right) + \left(\frac{V_A - V_o}{R_f}\right) + I_1 = 0$$

or

$$\frac{V_2 - V_A}{R_2} + \frac{V_A - V_o}{R_f} = 0 \qquad\qquad (4.26)$$

Writing Kirchhoff's Current Law at node 'B' yields

$$\left(\frac{V_1 - V_B}{R_1}\right) + \left(\frac{V_B - 0}{R_3}\right) + I_2 = 0$$

$$\left(\frac{V_1 - V_B}{R_1}\right) + \frac{V_B}{R_3} = 0$$

Substitute equation (4.25) here:

$$\frac{V_1 - V_A}{R_1} + \frac{V_A}{R_3} = 0$$

$$\frac{V_1 - V_A}{R_1} = -\frac{V_A}{R_3}$$

$$R_3(V_1 - V_A) = -V_A R_1$$

$$R_3 V_1 = V_A R_3 - V_A R_1$$

$$R_3 V_1 = V_A(R_3 - R_1)$$

or

$$V_A = \frac{R_3}{R_3 - R_1} V_1 \qquad (4.27)$$

Substitute equation (4.27) in equation (4.26),

$$\frac{V_2 - \dfrac{R_3}{R_3 - R_1} V_1}{R_2} + \frac{\dfrac{R_3}{R_3 - R_1} V_1 - V_o}{R_f} = 0$$

$$\frac{V_2 - \dfrac{R_3 V_1}{R_3 - R_1}}{R_2} = +\frac{V_o - \dfrac{R_3}{R_3 - R_1} V_1}{R_f}$$

$$R_f V_2 - \frac{R_3 R_f}{R_3 - R_1} V_1 = R_2 V_o - \frac{R_3}{R_3 - R_1} R_2 V_1$$

It is standard practice to take $R_1 = R_2$ and $R_3 = R_f$

$$R_f V_2 - \frac{R_f^{\,2}}{R_f - R_1} V_1 = R_1 V_o - \frac{R_f R_1}{R_f - R_1} V_1$$

$$R_1 V_o = R_f V_2 + \frac{R_f R_1}{R_f - R_1} V_1 - \frac{R_f^{\,2}}{R_f - R_1} V_1$$

$$= R_f V_2 + \frac{R_f V_1}{R_f - R_1} (R_1 - R_f)$$

$$= R_f V_2 - \frac{R_f V_1}{R_f - R_1} (R_f - R_1)$$

$$R_1 V_o = R_f V_2 - R_f V_1$$

$$R_1 V_o = R_f (V_2 - V_1)$$

$$V_o = \frac{R_f}{R_1} (V_2 - V_1) \tag{4.28}$$

Therefore, output is the amplified version of the difference between the two input terminals.

4.5. INSTRUMENTATION AMPLIFIER

An instrumentation amplifier is used to measure and control physical quantities such as temperature, humidity, light intensity, and waterflow.

The features needed for an instrumentation amplifier are

(a) high gain accuracy,

(b) high CMRR,

(c) high gain stability with very low temperature variation,

(d) low DC offset voltage, and

(e) low output impedance.

A basic instrumentation amplifier is a simple difference amplifier. In practice, a three-stage difference amplifier is used as an instrumentation amplifier, as shown in Figure 4.8.

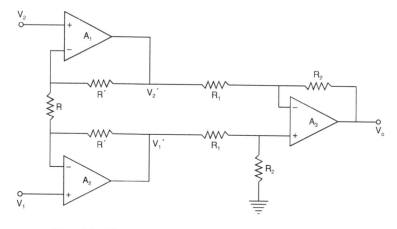

Fig. 4.8. Three-op-amp instrumentation amplifier.

For this circuit, the output V_o can be derived as

$$V_o = \frac{R_2}{R_1}\left(1 + \frac{2R'}{R}\right)(V_1 - V_2) \qquad (4.29)$$

By varying resistance R, the gain can be varied.

The practical circuit to measure the variation in a physical quantity is shown in Figure 4.9.

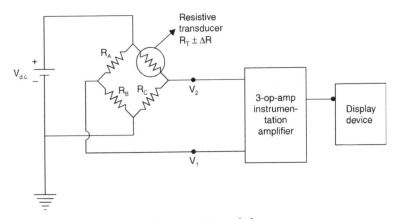

Fig. 4.9. Implementation of three-op-amp
instrumentation amplifer.

Initially, the bridge is balanced to get the output to be zero. As the physical quantity varies, the resistance in one arm is varied, therefore $V_1 \neq V_2$ and as a consequence, the output shown in the display device is not equal to zero.

4.6. TRANSRESISTANCE AMPLIFIER AND TRANSCONDUCTANCE AMPLIFIER

A transresistance amplifier converts an input of current to an output of voltage. It is also called a voltage to current converter or V to I converter. It is called transresistance because the efficiency of the amplifier is measured in units of resistance.

The analysis will be done assuming ideal op-amp characteristics. The current entering the op-amp terminals is zero. Accordingly, the current coming from the source will essentially flow through R_f.

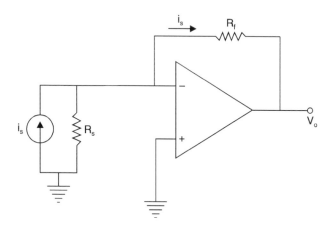

Fig. 4.10. Transresistance amplifier.

The voltage difference between the two input terminals is zero. Since the voltage at the non-inverting input terminal is zero, the voltage at the inverting input terminal is zero.

∴ Current through feedback resistance R_f can be calculated.

$$i_s = \frac{\text{Voltage difference}}{R_f}$$

$$= \left(\frac{0 - V_o}{R_f} \right)$$

$$-V_o = i_s R_f$$

$$V_o = -i_s R_f \tag{4.30}$$

Therefore, output voltage is proportional to input current.

A transconductance amplifier converts an input of voltage to an output of current. It is also called a current to voltage converter or I to V converter. It is called transconductance because the efficiency of the amplifier is measured in units of conductance.

Transconductance amplifiers are classified into two types. They are transconductance amplifiers with floating load and transconductance amplifiers with grounded load.

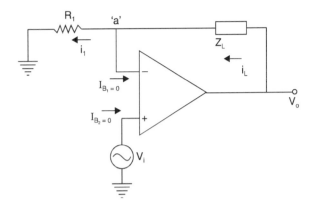

Fig. 4.11. Transconductance amplifier with floating load.

In the ideal op-amp,

$$V_a = V_i \qquad (4.31)$$

and $$I_{B1} = I_{B2} = 0 \qquad (4.32)$$

Writing Kirchhoff's Current Law at node 'a' yields

$$i_L + I_{B1} = i_1$$

$$i_L = i_1 \qquad (4.33)$$

Write the equation for current through R_1:

$$i_1 = \frac{V_a - 0}{R_1}$$

$$i_1 = \frac{V_a}{R_1}$$

Substitute equation (4.31) here:

$$i_1 = \frac{V_i}{R_1} \qquad (4.34)$$

Substitute equation (4.34) in (4.33),

$$i_L = \frac{V_i}{R_1} \qquad (4.35)$$

Here the output current is made proportional to input voltage.

4.7. MULTIPLIER AND DIVIDER

In general, a multiplier is shown with the basic schematic symbol in Figure 4.12.

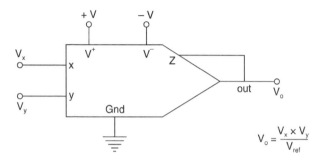

Fig. 4.12. Symbol for a multiplier.

The output voltage can be written as

$$V_o = \frac{V_x \times V_y}{V_{ref}}$$ (4.36)

where V_x and V_y are input signals

and V_{ref} is a reference voltage that is normally set to 10 volts. As long as $V_x < V_{ref}$ and $V_y < V_{ref}$, output will be less than V_{ref}.

If both inputs are positive, then the multiplier is called a one-quadrant multiplier. If one input is kept at a positive value and the other input is allowed to take either a positive or negative value, then it is called a two-quadrant multiplier. If both the inputs are allowed to take either positive or negative values, then it is called a four-quadrant multiplier.

There are some commercially available multiplier ICs and multiplier circuits can be constructed from op-amp ICs such as 741.

The applications of multipliers include frequency doubling, frequency shifting, phase angle detection, real power computation, and squaring signals.

A divider uses the multiplier in the feedback as shown in Figure 4.13.

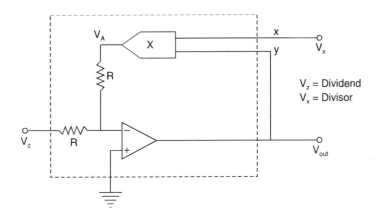

Fig. 4.13. Divider.

V_{out} can be written as

$$V_{out} = -V_{ref} \frac{V_z}{V_x} \tag{4.37}$$

The applications of dividers include taking square root and dividing one number by another.

4.8. DIFFERENTIATOR AND INTEGRATOR

The output of a differentiator, or differentiating amplifier, is the differentiated version of input given.

In an ideal op-amp, the voltage difference between the input terminals is zero. Since the voltage at the non-inverting input terminal is zero, the voltage at the inverting input terminal should also be zero.

$$\therefore \qquad\qquad V_N = 0 \tag{4.38}$$

The currents entering the op-amp input terminals are zero.

$$I_{B1} = I_{B2} = 0 \tag{4.39}$$

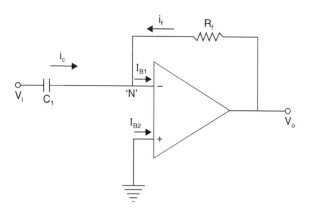

Fig. 4.14. Basic differentiator.

Writing Kirchhoff's Current Law at node 'N' yields

$$i_c + i_f = I_{B1}$$

Substitute equation (4.39):

$$i_c + i_f = 0$$

$$\therefore \qquad i_c = -i_f$$

or $$i_f = -i_c \qquad (4.40)$$

Write the current through the capacitor in terms of voltage:

$$i_c = C_1 \frac{d}{dt}(v_i - v_N)$$

Substitute equation (4.38):

$$i_c = C_1 \frac{dV_i}{dt} \qquad (4.41)$$

Write the current through the feedback resistor in terms of voltage:

$$i_f = \frac{(V_o - V_N)}{R_f}$$

Substitute equation (4.38):

$$i_f = \frac{V_o}{R_f}$$

$$V_o = i_f R_f$$

Substitute equation (4.40):

$$V_o = -i_c R_f$$

Substitute equation (4.41):

$$V_o = -C_1 \left(\frac{dV_i}{dt} \right) R_f$$

$$V_o = -C_1 R_f \frac{dV_i}{dt} \tag{4.42}$$

Equation (4.42) shows output voltage is proportional to derivative of input voltage

$$V_o \propto \frac{dV_i}{dt}$$

To have stability and to reduce noise, a resistor R_1 is placed in series with C_1 and a capacitor C_f is placed in parallel with R_f in a practical differentiator circuit.

The output of an integrator, or integrating amplifier, is the integrated version of the input.

The circuit for an integrator is the same as that of a differentiator, except the positions of the capacitor and resistor are switched.

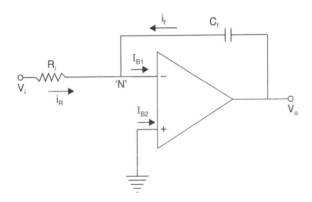

Fig. 4.15. Basic integrator.

In an ideal op-amp, the voltage difference between the input terminals is zero. Since the voltage at the non-inverting input terminal is zero, the voltage at the inverting input terminal is zero.

$$\therefore \qquad\qquad V_N = 0 \qquad\qquad (4.43)$$

The currents entering the op-amp input terminals are zero.

$$I_{B1} = I_{B2} = 0 \qquad\qquad (4.44)$$

Writing Kirchhoff's Current Law at node 'N' yields

$$i_R + i_f = I_{B1}$$

Substitute equation (4.44):

$$i_R + i_f = 0$$

$$\therefore \qquad\qquad i_f = -i_R \qquad\qquad (4.45)$$

Write the current through the resistor in terms of input voltage:

$$i_R = \frac{V_i - V_N}{R_1}$$

Substitute equation (4.43):

$$i_R = \frac{V_i}{R_1} \qquad\qquad (4.46)$$

Write the feedback current in terms of voltage:

$$i_f = C_f \frac{d(V_o - V_N)}{dt}$$

Substitute equation (4.43):

$$i_f = C_f \frac{dV_o}{dt} \qquad\qquad (4.47)$$

Substitute equations (4.46) and (4.47) in equation (4.45):

$$C_f \frac{dV_o}{dt} = -\frac{V_i}{R_1}$$

$$\frac{dV_o}{dt} = -\frac{V_i}{C_f R_1}$$

$$V_o = -\frac{1}{R_1 C_f} \int_0^t V_i \, dt + V_0(0) \qquad (4.48)$$

$R_1 C_f$ is the time constant.

Since $V_0(0)$ is the initial output voltage, when $t = 0$, it is a constant.

$$V_0 \propto \int_0^t V_i \, dt + \text{ constant}$$

Since the right hand side of the above equation contains 2 terms, of which one is constant, V_0 varies as the first term varies.

Therefore, output voltage varies proportionally to the variations of the integral of the input.

To have stability and to reduce noise, a resistor R_f is placed in parallel with the feedback capacitor C_f in a practical integrator circuit.

4.9. OP-AMP APPLICATIONS USING DIODES

The major limitation of an ordinary diode, which leads to the principle of using diodes with op-amps, is that an ordinary diode cannot rectify voltages below V_r, the cut-in voltage of the diode. Therefore, ordinary diodes cannot be used to cut off the circuit for rectification of voltages below cut-off voltages, which typically range from 0.6 V to 0.7 V.

The important circuits using diodes are rectifiers, clippers, clampers, and peak value storage devices.

Half-Wave Rectifier

The circuit for a half-wave rectifier is shown in Figure 4.16.

When the input signal is positive, i.e., $V_i > 0$, D_1 is forward biased and D_1 conducts. D_2 is reverse biased because $V_N = -V_r$ and D_2 does not conduct. Therefore, no current flows through R_f and output voltage $V_o = 0$.

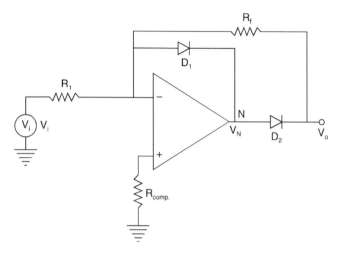

Fig. 4.16. Half-wave rectifier.

When the input signal is negative, *i.e.*, $V_i < 0$, D_1 is reverse biased and D_1 does not conduct. Therefore, D_2 conducts and the circuit behaves like an inverter causing the output voltage to become positive.

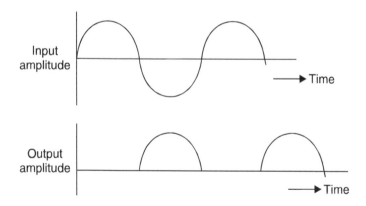

Fig. 4.17. Input-output relationship for a half-wave rectifier.

Full-Wave Rectifier

The circuit for a full-wave rectifier, or absolute value circuit or modulus value circuit is shown in Figure 4.18.

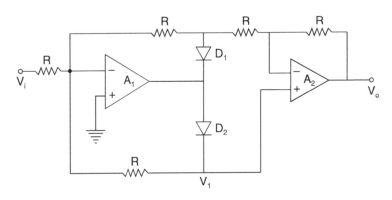

Fig. 4.18. Full-wave rectifier.

When the input voltage V_i is positive, D_1 is forward biased, hence D_1 conducts and D_1 is in an 'ON' state, and D_2 is reverse biased, hence D_2 does not conduct and D_2 is in an 'OFF' state. The equivalent circuit can be drawn as shown in Figure 4.19.

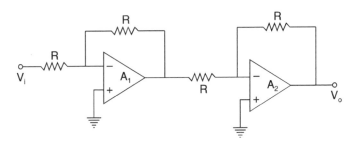

Fig. 4.19. Equivalent circuit of a full-wave rectifier for $V_i > 0$.

A_1 and A_2 act as inverters with a gain of $\dfrac{R_f}{R_i} = \dfrac{R}{R} = 1$.

Therefore, output of the circuit is the same as the input.

$$V_o = V_i$$

When input V_i is negative, D_1 is in an 'OFF' state and D_2 is in an 'ON' state. In this case, the equivalent circuit can be drawn as shown in Figure 4.20.

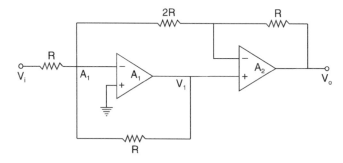

Fig. 4.20. Equivalent circuit for a full-wave rectifier for $V_i < 0$.

Writing Kirchhoff's Current Law at the input node and using ideal op-amp parameters for A_1 and A_2, the sum of the currents entering node 'A' should be zero.

$$\frac{V_i}{R} + \frac{V_1}{R} + \frac{V_1}{2R} = 0$$

$$\therefore \qquad 2V_i + V_1 + 2V_1 = 0$$

$$3V_1 = -2V_i$$

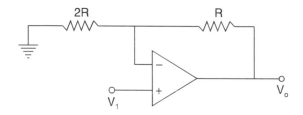

Fig. 4.21. Simplified circuit of Figure 4.20.

$$V_1 = -\frac{2}{3}V_i$$

$$V_o = \left(1 + \frac{R}{2R}\right)\left(-\frac{2}{3}V_i\right)$$

$$= -V_i$$

$\therefore$ For negative input, output = + ve.

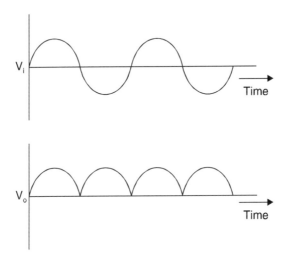

Fig. 4.22. Input-output relationship for a full-wave rectifier.

Clipper

A clipper is a circuit in which a certain portion of the input is clipped off, *i.e.*, made constant.

If the clipping is done on the positive side of V_{ref}, then it is called a positive clipper circuit. If the clipping is done on the negative side of V_{ref}, then it is called a negative clipper circuit. A positive clipper is shown in Figure 4.23.

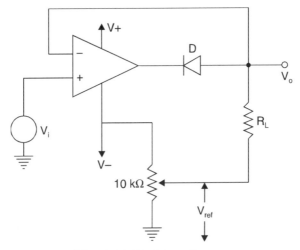

Fig. 4.23. Positive clipper.

Clipping level is determined by $V_{ref.}$. Whenever $V_i < V_{ref}$, diode D is forward biased and D conducts. Therefore, the circuit acts as a voltage follower circuit.

Whenever $V_i > V_{ref}$, diode D is reverse biased and D does not conduct. Therefore, output voltage $V_o = V_{ref}$.

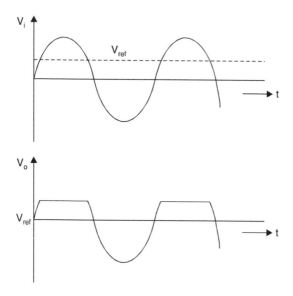

Fig. 4.24. Input-output relationship for a positive clipper.

V_{ref} can be changed by changing the potentiometer position.

By reversing the direction of diode D, a negative clipper circuit can be constructed.

Clamper

A clamper circuit, or DC restorer or restorer or DC inserter, adds a desired DC level to the signal. The typical clamper circuit is shown in Figure. 4.25.

If the added DC level is positive, it is called a positive clamper. If the added DC voltage is negative, it is called a negative clamper.

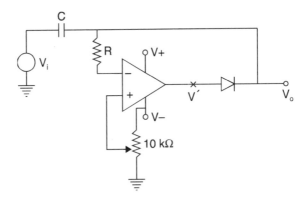

Fig. 4.25. Clamper circuit.

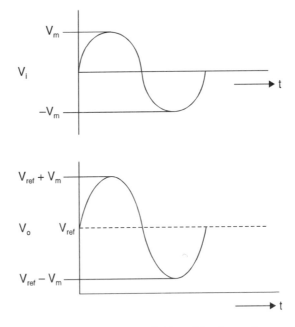

Fig. 4.26. Input-output relationship for a clamper.

4.10. COMPARATOR

A comparator compares a signal voltage applied at the input of an op-amp with a known reference voltage V_{ref} given at the other input. It is an open-loop operation, *i.e.*, there is no feedback path in the case of a comparator.

Comparators can be classified into two types, namely non-inverting and inverting. In both cases, V_{ref} may be either positive or negative. The output of a comparator can be in one of two states, $+V_{sat}$ or $-V_{sat}$. If the voltage at inverting input terminal is greater than the voltage at non-inverting input terminals then $V_o = -V_{sat}$, otherwise $V_o = +V_{sat}$.

Non-Inverting Comparator

The signal input is given to the non-inverting input terminal and the reference voltage is given to the inverting input terminal.

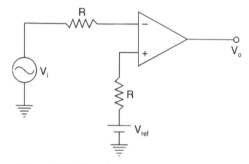

Fig. 4.27. Non-inverting comparator.

As long as input $V_i < V_{ref}$, the output will be $-V_{sat}$.

If input $V_i > V_{ref}$, the output voltage will be $+V_{sat}$.

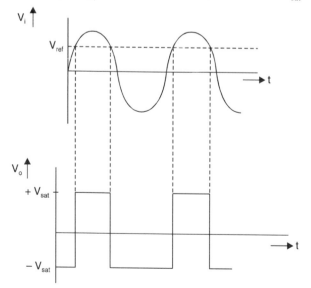

Fig. 4.28. Input-output for non-inverting comparator.

In a practical comparator, V_{ref} is obtained by using a potentiometer between two power supply terminals. By adjusting the potentiometer, the V_{ref} voltage is varied.

Inverting Comparator

The signal input is given to the inverting input terminal and the reference voltage is given to the non-inverting input terminal.

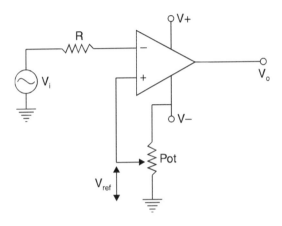

Fig. 4.29. Inverting comparator.

Whenever input voltage $V_i > V_{ref}$, the output voltage will be $+ V_{sat}$.

Whenever input voltage $V_i < V_{ref}$, the output voltage will be $- V_{sat}$. Input and output waveforms can be drawn, as was done for non-inverting comparators in Figure 4.28.

4.11. WINDOW DETECTOR

A window detector is a special type of comparator. In this case, many comparators will be working in parallel, with different reference voltages. By noting the outputs from the different comparators, the range which contains the input can be noted.

A typical window detector circuit is shown in Figure 4.30.

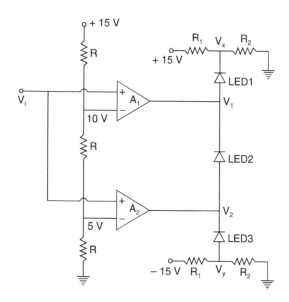

Fig. 4.30. Window detector circuit.

In this example, two non-inverting comparators are used. The total voltage, 15 V, is divided into three equal voltage steps, with the help of three equal resistances R. Therefore, the voltage above one resistance will be + 5 V and the voltage above two resistances will be + 10 V.

V_x can be calculated from R_1 and R_2 and can be noted as a positive voltage. Similarly V_y can be noted as a negative voltage.

If input V_i is less than 5 V, A_2 produces a output of $-V_{sat}$ and A_1 also produces a output of $-V_{sat}$. In this state LED3 is forward biased and LED2 is not forward biased to the sufficient extent, and LED1 is reverse biased. Since forward biased LEDs will emit light, LED3 will glow and the other LEDs will not.

If input V_i is between 5 V and 10 V, A_2 produces an output of $+V_{sat}$ and A_1 produces an output of $-V_{sat}$. Under this condition LED3 is reverse biased, LED2 is forward biased, and LED1 is reverse biased. Therefore, LED2 will glow and LED1 and LED3 will not.

If input V_i is greater than 10 V, A_2 produces an output of $+V_{sat}$ and A_1 produces an output of $+V_{sat}$. This condition forward biases LED1 and reverse biases the other two LEDs. Therefore, LED1 will glow and the other LEDs will not.

By noting the three output LEDs, we can note the input voltage range. Since the entire input range has been divided into a stipulated window, this circuit is known as window detector.

4.12. MULTIPLIER AS MIXER

The basic principle of a multiplier is shown in section 4.7. A multiplier will produce an output that is proportional to the product of the two inputs.

$$V_o = \frac{V_x \times V_y}{V_{ref}} \tag{4.49}$$

Let V_x and V_y be sine waves of different frequencies.

$$V_x = E_x \times \cos(\omega_x t + \theta_x) \tag{4.50}$$

$$V_y = E_y \times \cos(\omega_y t + \theta_y) \tag{4.51}$$

Substitute equations (4.50) and (4.51) in equation (4.49):

$$V_o = \frac{E_x \times \cos(\omega_x t + \theta_x) \times E_y \times \cos(\omega_y t + \theta_y)}{V_{ref}}$$

$$= \frac{E_x \times E_y}{V_{ref}} \times \cos(\omega_x t + \theta_x) \cos(\omega_y t + \theta_y)$$

Assume $\theta_x = \theta_y = 0$, for simplicity:

$$V_o = \frac{E_x \times E_y}{V_{ref}} \cos \omega_x t \times \cos \omega_y t \tag{4.52}$$

Substitute from basic mathematics:

$$\cos C \cos D = \frac{\cos(C+D) + \cos(C-D)}{2} \tag{4.53}$$

Assume cos terms in equation (4.52) in the form as shown in equation (4.53) and apply equation (4.53):

$$V_o = \frac{E_x \times E_y}{V_{ref}} \times \left[\frac{1}{2}\right] [\cos\{(\omega_x + \omega_y)t\} + \cos\{(\omega_x - \omega_y)t\}]$$

$$V_o = \frac{E_x \times E_y}{2V_{ref}} \left[\cos\{(\omega_x + \omega_y)t\} + \cos\{(\omega_x - \omega_y)t\}\right]$$

$$V_o = \frac{E_x \times E_y}{2V_{ref}} \cos(\omega_x + \omega_y)t + \frac{E_x \times E_y}{2V_{ref}} \cos(\omega_x - \omega_y)t$$

$$(4.54)$$

Equation (4.54) has two terms on the right hand side. The first term has a frequency equal to the sum of the two input frequencies and the second term has a frequency equal to the difference between the two input frequencies. Therefore, the output of a multiplier produces sum and difference terms, which must come from a mixer. Thus by giving sinusoidal inputs of different frequencies to a multiplier circuit, a multiplier can be used as a mixer.

4.13. ACTIVE HIGH-PASS AND LOW-PASS FILTERS

A filter, in general, may be defined as a frequency-selective electronic circuit. A filter circuit allows some frequencies to pass through it and attenuates others.

The simplest way to make a filter is to use a resistor, an inductor, and a capacitor. This filter is called a passive filter because it does not contain any active component. If there are some active components in a filter circuit, then it is called an active filter. This discussion will pertain to active filters using op-amps as active components.

Active filters can be classified depending on the type of filtering action, *e.g.,* low-pass filter, high-pass filter, etc. Active filters can also be classified depending on the order of the filter, *i.e.,* the number of RC pole pairs, into first-order filter or second-order filter.

Some examples of high-pass and low-pass filters will be discussed here.

First-Order Low-Pass Filter

The basic configuration used is a non-inverting amplifier.

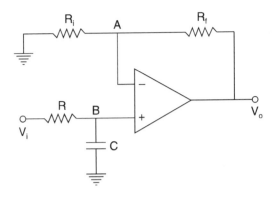

Fig. 4.31. First-order low-pass filter.

Assuming ideal op-amp conditions, the currents entering the op-amp terminals are zero.

At the non-inverting input terminal, the current flowing through resistor R should be equal to the current flowing through capacitor C.

∴ The voltage drop across the capacitor, *i.e.*, voltage at non-inverting input terminal is written:

$$V_B(S) = \frac{\left(+ \dfrac{1}{SC} \right)}{\left(R + \dfrac{1}{SC} \right)} \, V_i(S)$$

where V_B is the input voltage.

$$\therefore \qquad \frac{V_B(S)}{V_i(S)} = \frac{1}{RCS + 1} \tag{4.55}$$

The voltage gain for the non-inverting amplifier is

$$A_0 = 1 + \frac{R_f}{R_i} = \frac{V_o(S)}{V_B(S)} \tag{4.56}$$

Therefore, overall transfer function of the network can be written:

$$H(S) = \frac{V_o(S)}{V_i(S)} = \frac{V_o(S)}{V_B(S)} \times \frac{V_B(S)}{V_i(S)}$$

Substitute equations (4.55) and (4.56):

$$H(S) = \frac{1}{RCS + 1} \left(1 + \frac{R_f}{R_i} \right)$$

$$H(S) = \frac{1}{RCS + 1} \times (A_0)$$

Substitute $RC = \dfrac{1}{\omega_h}$:

$$H(S) = \frac{A_0}{S / \omega_h + 1}$$

$$H(S) = \frac{A_0 \omega_h}{S + \omega_h} \tag{4.57}$$

To determine frequency response, substitute $S = j\omega$:

$$H(j\omega) = \frac{A_0 \times \omega_h}{j\omega + \omega_h}$$

$$= \frac{A_0}{\dfrac{j\omega}{\omega_h} + 1}$$

$$H(j\omega) = \frac{A_0}{1 + j \left(\dfrac{f}{f_h} \right)} \tag{4.58}$$

By substituting increasing values for f, the response can be noted as a low-pass filter because as frequency increases, the transfer function decreases in value.

f_h is equal to a 3 dB cut-off frequency because at $f = f_h$, the transfer function gives an output which is 3dB less than the maximum magnitude. This cut-off frequency is also referred to as half-power frequency because the power is equal to half the maximum power. This frequency is also called corner frequency or break frequency.

The frequency range where the gain of the transfer function is between 0 and 3 dB gain is called the stop band. For the low-pass filter, the stop band will be from f_h to $\propto$ frequency.

The frequency range where the gain of the transfer function is between maximum gain and 3 dB gain is called the pass band. For the low-pass filter, the pass band will be from 0 Hz to f_h Hz.

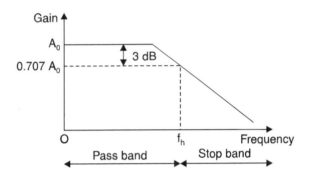

Fig. 4.32. Frequency response of first-order low-pass filter.

Second-Order Low-Pass Filter

This structure should have two RC pole pairs.

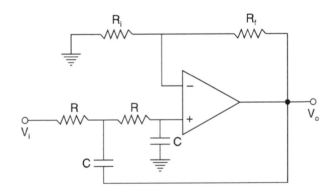

Fig. 4.33. Second-order low-pass filter.

The transfer function of this structure is

$$H(S) = \frac{A_0}{S^2 C^2 R^2 + SCR(3 - A_0) + 1} \qquad (4.59)$$

By inserting different values of frequencies and calculating the transfer functions at different frequencies, the structure could be proved to act as a low-pass filter. For a low-pass filter, the transfer function will be A_0 at 0 frequency and, as frequency increases, the gain will decrease.

The cut-off frequency can be defined:

$$f_h = \frac{1}{2\pi RC} \tag{4.60}$$

Equation (4.59) can be rewritten:

$$H(S) = \frac{A_0 \times \omega_h^2}{S^2 + \alpha\omega_h S + \omega_h^2} \tag{4.61}$$

where, A_0 is the gain of non-inverting amplifier as given by equation (4.56).

α is defined as a damping coefficient.

It can be shown that

$$\alpha = 3 - A_0 \tag{4.62}$$

The magnitude response of a second-order low pass filter can be written:

$$20 \log \mid H(j\omega) \mid = 20 \log \frac{A_0}{\sqrt{1 + \left(\dfrac{\omega}{\omega_h}\right)^4}} \tag{4.63}$$

The magnitude response of a general nth-order low pass filter can be written:

$$20 \log \mid H(j\omega) \mid = 20 \log \frac{A_0}{\sqrt{1 + \left(\dfrac{\omega}{\omega_h}\right)^{2n}}} \tag{4.64}$$

Higher-order low-pass filters can be built by cascading first-order and second-order low-pass filters.

As the order of the filter increases, the higher the roll-off rate occurs in the stop band, the quicker the response falls to zero gain.

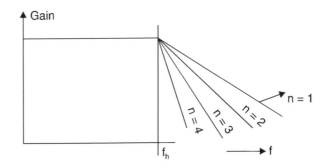

Fig. 4.34. Frequency response for different order filters.

The roll-off rate for an nth-order filter is given as $-n \times 20$ dB/decade in the stop band.

First-Order High-Pass Filter

A high-pass filter is the complement of a low-pass filter. The construction of any high-pass filter is the same as that of a low-pass filter except that the resistor and capacitor are switched.

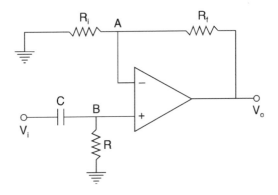

Fig. 4.35. First-order high-pass filter.

It can be proved that

$$H(S) = \frac{A_0 S}{S + \omega_l} \qquad (4.65)$$

where $\quad \omega_l = \dfrac{1}{RC} \qquad (4.66)$

The frequency response can be determined by calculating the magnitude of transfer function at many frequencies.

$$H(j\omega) = \frac{A_0}{1 + j\left(\dfrac{f_l}{f}\right)} \qquad (4.67)$$

f_l is called the 3 dB cut-off frequency, half-power frequency, corner frequency, or break frequency.

For the high-pass filter, the pass band lies from f_l Hz to $\propto$ Hz and the stop band lies from 0 Hz to f_l Hz.

$$f_l = \frac{1}{2\pi RC} \qquad (4.68)$$

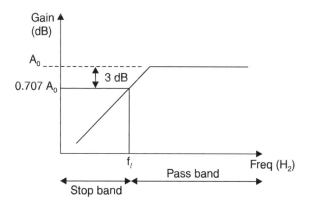

Fig. 4.36. Frequency response of a
first-order high-pass filter.

Second-Order High-Pass Filter

This structure can be made by switching the positions of resistor R and capacitor C in the second-order low-pass filter.

It can be derived that

$$H(S) = \frac{A_0 S^2}{S^2 + (3 - A_0)\omega_l S + \omega_l^{\,2}} \qquad (4.69)$$

and
$$\omega_l = \frac{1}{RC}$$

or
$$f_l = \frac{1}{2\pi RC}$$
(4.70)

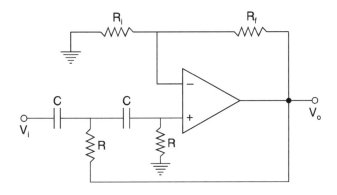

Fig. 4.37. Second-order high-pass filter.

$$20 \log | H(j\omega) | = \frac{A_0}{\sqrt{1 + \left(\dfrac{f_1}{f}\right)^4}}$$
(4.71)

For a general nth-order high-pass filter, the magnitude of transfer function can be written:

$$20 \log | H(j\omega) | = \frac{A_0}{\sqrt{1 + \left(\dfrac{f_1}{f}\right)^{2n}}}$$
(4.72)

As the order of filter varies, the roll-off rate varies, as discussed for the low-pass filter. The only difference is roll-off occurs before the cut-off frequency for the high-pass filter. Band-pass and bank-reject filters can be constructed from high-pass and low-pass filters by selecting proper cut-off frequencies for low-pass and high-pass filters.

5

Phase Locked Loops

Phase Locked Loops (PLL) are very important blocks in a linear system. They are used for synchronization in radar applications, satellite communication, airborne navigational systems, frequency modulation communication systems, and more.

5.1. BASIC PRINCIPLES

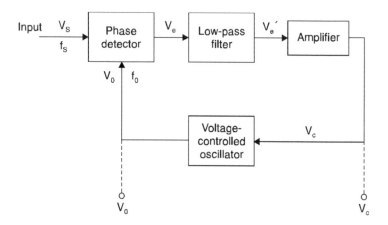

Fig. 5.1. Internal diagram of a PLL.

An input signal is given as one input of the phase detector. The other input to the phase detector is the output of a voltage-controlled oscillator. The output of the phase detector V_e becomes the input of a low-pass filter. The output of that filter V_e' becomes the input of an amplifier. The output of the amplifier V_c becomes the control voltage for the voltage-controlled oscillator.

The voltage controlled oscillator (VCO) mentioned here is nothing but a free-running multivibrator, *i.e.,* astable multivibrator. Its normal frequency of oscillation is f_0 and its normal voltage is V_0. By applying a DC voltage as control voltage for the VCO, the frequency of its oscillation can be shifted in both directions. The frequency deviation of this device is proportional to the DC control voltage, hence its name.

Operation. Let the assumption be made that the input signal has an amplitude V_s and a frequency f_s.

A phase detector compares two input signal frequencies and the phase of them. If they differ, it will produce an output error voltage, otherwise not. If $f_s \neq f_0$, then V_e is produced. In general, the phase detector will produce output having two frequency components $(f_s + f_0)$ and $(f_s - f_0)$. The purpose of the low-pass filter is to absorb the sum frequency generated and allow the difference frequency to propagate as output V_e'.

The V_e' signal is amplified and V_c is given as the control voltage for the VCO. V_c changes the output frequency of the VCO in such a way that the difference between f_s and f_0 is reduced. Once the frequency of the VCO is moving toward the input signal frequency, it is said that *capturing action* has started to take place.

If $f_s \neq f_{0\,\text{new}}$, then the same set of actions take place, until $f_s = f_0$. Once the frequency of the VCO is equal to the signal frequency, then it is said that *locking action* has started.

Once locked, a change in the input signal frequency, is reflected soon in the output frequencyof the VCO. The frequency of the VCO is always kept the same as the input signal frequency.

In general, the PLL is said to undergo three states.

(*a*) **Free-running state.** When there is no input signal, the VCO oscillates with its free-running oscillating frequency. This is referred to as free-running state.

(*b*) **Capturing state.** As soon as the input signal is applied, the VCO frequency is changed according to the input frequency. This state is referred to as capturing state.

(c) **Locked state.** In this state, the VCO and the input signal are locked, *i.e.,* have the same frequency of oscillation. Whenever the input signal varies in frequency, the same is reflected in the VCO frequency.

There are some parameters to be defined in terms of PLL.

(a) **Lock-in or Tracking Range.** Defined as the range of frequencies over which the PLL can maintain the lock with the incoming signal, this parameter comes into play in the locked state.

(b) **Capture Range.** Defined as the range of frequencies over which the PLL can create a lock with an incoming signal, this parameter comes into play when the PLL is in the free-running state.

Tracking range will always be greater than capture range.

(c) **Pull-in Time.** The time needed for the PLL to establish a lock with an applied input signal depends on the difference between the input signal and the free-running frequencies. This time is referred to as pull-in time.

5.2. VOLTAGE-CONTROLLED OSCILLATOR

A typical example of a VCO is IC NE 566. The pin diagram is given in Figure 5.2.

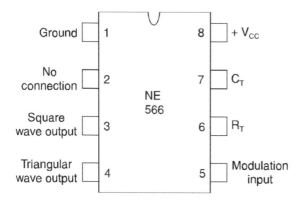

Fig. 5.2. Pin diagram of IC NE 566.

The equivalent block diagram of IC NE 566 is shown in Figure 5.3.

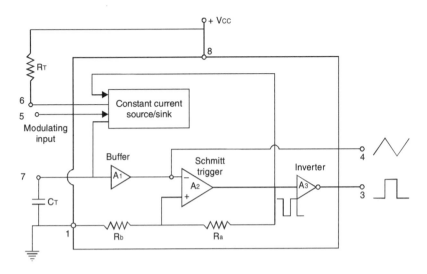

Fig. 5.3. Internal diagram of IC 566.

The constant current source/sink supplies or absorbs a constant current. The capacitor connected between pins 7 and 1, *i.e.*, C_T, is charged or discharged linearly by this constant current source/sink. The value of the constant current source can be changed by varying the modulating input voltage V_C applied to pin 5.

The voltage across C_T is applied to the inverting input terminal of Schmitt trigger A_2 through the buffer A_1. The Schmitt trigger is designed in such a way that the output voltage changes whenever the input voltage crosses $0.5 \ V_{CC}$ and V_{CC}.

By selecting the proper values for R_a and R_b, the oscillating voltage for the capacitor can be changed. If $R_a = R_b$,

$$\beta = \frac{R_a}{R_a + R_b} = 0.5.$$ This leads to the fact that whenever the

capacitor voltage crosses the limits $0.25 \ V_{CC}$ and $0.5 \ V_{CC}$, there is a transition in the output.

Consider the following case. The voltage at capacitor C_T exceeds $0.5 \ V_{CC}$ during the charging action of the capacitor. The output of the Schmitt trigger should go low *i.e.*, $0.5 \ V_{CC}$, because the capacitor voltage is given to the inverting input terminal.

Since the output of the Schmitt trigger is 0.5 V_{CC}, the capacitor tries to discharge. When the capacitor attains a voltage lower than 0.25 V_{CC}, voltage of the non-inverting input terminal of the op-amp becomes higher than the voltage at the inverting input terminal. Hence, the output of the Schmitt trigger changes to V_{CC} and the capacitor starts charging toward V_{CC}.

Since the capacitor is charged and discharged through a constant current source/sink, the time taken for charging the capacitor from 0.25 V_{CC} to 0.5 V_{CC} is equal to the time taken for the capacitor to discharge from 0.5 V_{CC} to 0.25 V_{CC}. Since the rise time is equal to the fall time, the voltage across the capacitors appears as a triangular wave, which appears at pin 4.

The Schmitt trigger output varies between 0.5 V_{CC} and V_{CC} and also has the time$_{on}$ = time$_{off}$. Hence, a square wave appears at the output of the Schmitt trigger A_2. But at pin 3, we have the output, which is an inverted version of the output of A_2. The output at pin 3 is also a square wave, but an inverted version of the output of the Schmitt trigger.

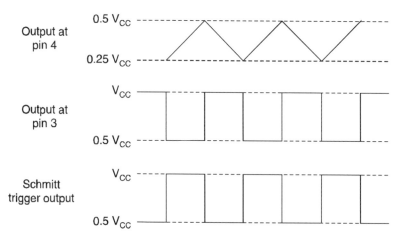

Fig. 5.4. Waveforms in IC NE 566.

As already stated, the VCO is a free-running multivibrator whose oscillating frequency can be changed by varying the modulating voltage.

As seen in previous paragraphs, the output at pin 3 is a square wave which goes on continuously. Therefore, this action is that of a free-running multivibrator.

By varying the modulating voltage at pin 5, the constant current source/sink varies its current supply/absorption capacity and the time taken for the capacitor to charge or discharge varies. Therefore, the time period of the waveforms varies.

It can be deduced that the frequency of oscillation is given by

$$f_0 = \frac{2(V_{CC} - v_c)}{C_T R_T V_{CC}} \tag{5.1}$$

A voltage to frequency conversion factor (K_v) is defined as the ratio of change in oscillating frequency of the VCO to the change in modulation voltage which caused this change in oscillating frequency.

$$K_V = \frac{\Delta f_0}{\Delta v_c} \tag{5.2}$$

5.3. LOW-PASS FILTER

The purpose of the low-pass filter is to separate the difference frequency from the sum frequency which is rejected. The low-pass filter (LPF) can be either passive or active.

The important thing to note is that if the bandwidth of the LPF is reduced, the capture range of the PLL is decreased and the response time is increased.

5.4. MONOLITHIC PLL

Some monolithic PLLs are available commercially. Members of a series differ mainly in power requirements and operating frequency range.

IC PLL NE/SE 565

This is a PLL IC, commercially available as a 14-pin DIP or a 10-pin metal can package.

The salient feature of this PLL is that the VCO output is not connected to phase comparator internally; the user has to connect it.

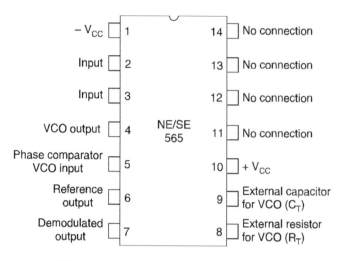

Fig. 5.5. Pin diagram of IC NE/SE 565.

The free-running frequency of voltage controlled oscillator is

$$f_0 = \frac{1}{4R_T C_T} \text{ Hz} \tag{5.3}$$

The important parameters of IC NE 565 are as follows:

Operating frequency range: 0.001 Hz to 500 kHz

Operating voltage range: ± 6 V to ± 12 V

Triangle wave amplitude: 2.4 V_{PP} at ± 6 V supply

Square wave amplitude: 5.4 V_{PP} at ± 6 V supply.

Lock-in range can be derived as

$$\Delta f_L = \pm \frac{7.8 f_0}{V} \tag{5.4}$$

where f_0 is given by equation (5.3) and V is given by

$$V = + V_{cc} - (- V_{cc}) \tag{5.5}$$

Capture range can be derived as

$$\Delta f_c \simeq \pm \sqrt{f_1 \Delta f_L} \tag{5.6}$$

where $\qquad\qquad f_1 = \dfrac{1}{2\pi RC} \tag{5.7}$

is the 3 dB cut-off frequency of the low-pass filter present in the PLL.

Lock-in range will always be greater than capture range.

5.5. PLL APPLICATIONS

In general, the output from the PLL can be taken in two ways:

(a) as a voltage signal at the output of the amplifier, which will be proportional to the error signal, between the input signal frequency and the VCO free-running frequency.

(b) as a frequency signal at the output of the VCO, which is the approximated version of the input signal frequency.

All the applications of a PLL can be classified into one of these two categories.

The applications of type (a) correspond to *frequency discriminator* applications. Here the output, *i.e.*, the control voltage V_c, varies in accordance with the instantaneous value of the frequency of input signal. This sort of application can be used for demodulation of frequency-modulated signals.

The applications of type (b) correspond to frequency synthesis and clock and carrier recovery applications. The idea is to pass on one signal frequency and signals of other frequencies will be attenuated. For example, if the input signal consists of a range of frequencies, the PLL can be made to lock on only one of those frequencies. This application is useful for reconditioning or regenerating a desired frequency signal from a composite signal.

Typical applications of type (a) are demodulation of FM signals and FSK signals.

Typical applications of type (*b*) are frequency multiplication, frequency division, frequency translation, and demodulation of AM waves.

5.6. PLL-BASED CLOCK AND CARRIER RECOVERY

This is an example of an application of type (*b*). The input signal consists of many signals, in which the desired clock and carrier signals are present. If the PLL is locked to the clock signal, then the output of the VCO inside the PLL will be a square wave with the same frequency as the clock signal. This is external output. Thus, the required clock signal has been recovered from a mixture of signals.

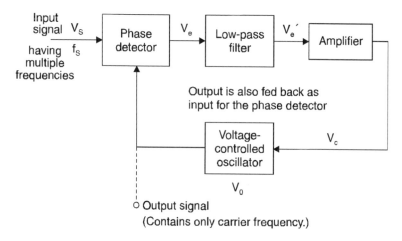

Fig. 5.6. PLL-based clock and carrier recovery.

6

555 Timer IC

IC NE/SE 555 is a highly stable device for generating accurate time delays. Commercially, this IC is available in 8-pin circular, TO-99 or 8-pin DIP, or 14-pin DIP packages.

The salient features of 555 timer ICs are

— Compatible with both TTL and CMOS logic families.

— The maximum load current can go up to 200 mA.

— The typical power supply voltage is from + 5 V to + 18 V.

— It can be used in a variety of applications such as oscillators, pulse generators, ramp generators, square wave generators, monoshot multivibrator burgular alarms, and traffic light controls.

The pin diagram of an 8-pin DIP 555 timer IC is given in Figure 6.1.

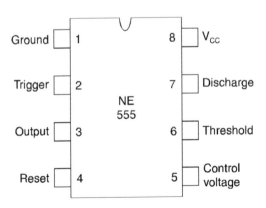

Fig. 6.1. Pin diagram of IC NE/SE 555.

74

6.1. TIMER INTERNAL DIAGRAM

The simplified form of internal diagram of IC 555 is shown in Figure 6.2. IC 555 consists of two comparators, a lower comparator and an upper comparator. It consists of a potential divider containing three 5 kΩ resistors and an RS flip flop, where the R input is connected to the upper comparator and the lower comparator output is the S input.

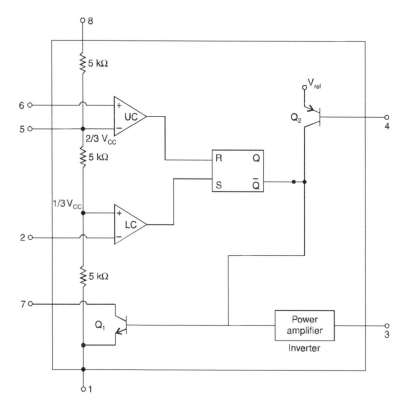

Fig. 6.2. Internal diagram of IC 555.

6.2. MONOSTABLE OPERATION

The connection diagram for monostable operation using IC 555 is shown in Figure 6.3. Pin 5 is connected to the ground through 0.01 µF, to keep pin 5 inactive, *i.e.,* the internal potential divider takes care of everything and no external modulation voltage is applied. Pin 4 is connected to $+V_{CC}$, the power supply. Pin 6 is connected to pin 7.

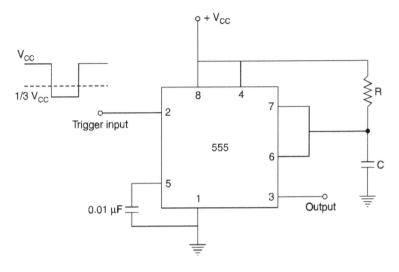

Fig. 6.3. Circuit for monostable multivibrator.

As long as the trigger input is not activated or the voltage at trigger input is greater than $1/3\ V_{CC}$, for the lower comparator, the voltage at the non-inverting input is lower than the voltage at the inverting input. Therefore, the output of the lower comparator is low, *i.e.*, $S = 0$, and the output of the upper comparator is high, *i.e.*, $R = 1$. For the RS flip flop, input $R = 1$ and $S = 0$. Hence the output $Q = 0$ and $\overline{Q} = 1$.

For transistor Q_1, the base voltage is high and the emitter voltage is low. Therefore, Q_1 conducts, in other words Q_1 is 'ON'. $\overline{Q} = 1$ and the inverter power amplifier changes the output, so the output at Pin 3 of IC 555 is 'LOW'.

The collector of transistor Q_1 is connected to the ground through the external capacitor C. Since transistor Q_1 is conducting, the voltage across the capacitor is clamped to the ground.

If the trigger input pin voltage is lower than $1/3\ V_{CC}$, for the lower comparator, the voltage at the non-inverting input is higher than the voltage at the inverting input. Therefore, the output of the lower comparator is high, $i.e.$, $S = 1$, and the output of the upper comparator is low. For RS flip flop, $S = 1$. Hence the output of the flip flop is $Q = 1$ and $\overline{Q} = 0$.

For the transistor Q_1, the base voltage and the emitter voltage are low. Therefore, the transistor is 'OFF' and there will not be any conduction by Q_1. The inverter power amplifier is given a low input, so the output at pin 3 of IC 555 will be high.

Since Q_1 is not conducting and capacitor C is not tied to the ground, the capacitor tries to charge to V_{CC}. Once the voltage across capacitor C crosses 2/3 V_{CC}, the voltage at the non-inverting input of the upper comparator is higher than the voltage at the inverting input of the upper comparator. Therefore, the output of the upper comparator is high, $i.e.$, $R = 1$. For the flip flop, the output will become $Q = 0$ and $\overline{Q} = 1$.

As explained previously, transistor Q_1 is 'ON' and capacitor C is tied to the ground and the output at Pin 3 of IC 555 is Low.

In summary, without application of any trigger pulses, the output at pin 3 was low all the time, meaning it was in a stable state.

With the application of a single pulse, the output became high and it remained high until the capacitor charged to 2/3 V_{CC}. Once the capacitor was charged to 2/3 V_{CC}, the output achieved the stable low state without needing any other pulse.

Time period T of oscillation, can be derived as

$$T = 1.1 \, RC \text{ seconds} \qquad (6.1)$$

If any reset pulse is applied before this period T and after the application of trigger pulse, output immediately achieves the stable low state. This happens because Q_2 is 'OFF' and Q_1 is 'ON'. Therefore, capacitor C is immediately discharged.

The typical applications of the monostable operation of timer ICs includes missing pulse detector, linear ramp generator, frequency divider, pulse width modulation, water level fill control, and touch switch.

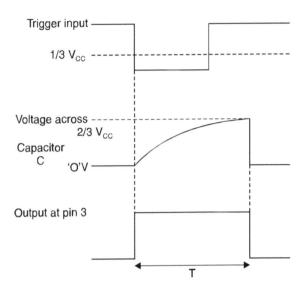

Fig. 6.4. Waveforms in monostable multivibrator.

Monostable indicates that of the two states of output, one is stable and the other is quasi-stable, *i.e.,* stable for a certain duration of time.

6.3. ASTABLE OPERATION

The name astable operation implies that there is no stable output state. The output will be switching between the two quasi-stable states continuously.

The timing resistor R used in monostable operation is split into two resistors and placed as shown in Figure 6.5.

When the voltage is applied, the capacitor is forced to charge through resistors R_A and R_B. Therefore, the time constant during which the capacitor is charging is $(R_A + R_B)C$. While the capacitor is charging, $R = 0$ and $S = 1$, (refer to the internal diagram of IC 555). Therefore, $\overline{Q} = 0$ and *output = high* during the charging of the capacitor. The capacitor will continue to charge as long as the voltage does not exceed 2/3 V_{CC}.

When the capacitor has attained a charge of 2/3 V_{CC} (to be precise, just greater than 2/3 V_{CC}), the upper comparator is triggered and $R = 1$ and $S = 0$. Therefore, $\overline{Q} = 1$ and *output* = *low*. Q_1 is 'ON' and the charge acquired by the capacitor will be discharged through Q_1. The resistance involved in the discharge path is R_B. Therefore, the time constant during which the capacitor discharges is $R_B C$. During this discharge period, *output* = *low*. This will continue until the voltage of the capacitor decreases to 1/3 V_{CC}.

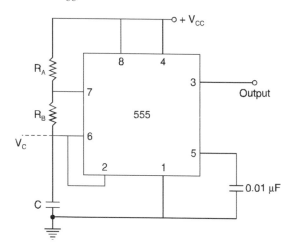

Fig. 6.5. Circuit for astable multivibrator.

When the capacitor has discharged to 1/3 V_{CC} (to be precise, just less than 1/3 V_{CC}), the lower comparator is triggered. $R = 0$ and $S = 1$, therefore *output* = *high*. Capacitor C is charged and these steps are repeated.

In summary, the capacitor is repeatedly charged and discharged between 1/3 V_{CC} and 2/3 V_{CC}. When the capacitor is charging, the output will remain high and when the capacitor is being discharged, the output will remain low. This can be drawn as shown in Figure 6.6.

It can be proved that

$$T_{High} = 0.69 \ (R_A + R_B)C \qquad\qquad (6.2)$$
$$T_{Low} = 0.69 \ R_B C \qquad\qquad\qquad (6.3)$$

Frequency of oscillation can be calculated as

$$f = \frac{1}{T} = \frac{1}{T_{High} + T_{Low}}$$

$$f = \frac{1.45}{(R_A + 2R_B)C} \tag{6.4}$$

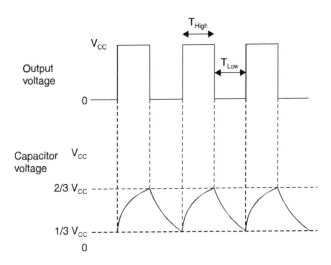

Fig. 6.6. Waveforms in astable multivibrator.

Duty cycle of square wave is defined as

$$D = \frac{T_{ON}}{T_{ON} + T_{OFF}} \tag{6.5}$$

Here T_{ON} indicates that Q_1 is 'ON' and $T_{ON} = T_{Low}$. Similarly, T_{OFF} indicates that Q_1 is 'OFF' and $T_{OFF} = T_{High}$. Substitute these in equation (6.5):

$$D = \frac{T_{Low}}{T_{Low} + T_{High}}$$

By substituting equations (6.2) and (6.3), it can be proved that

$$D = \frac{R_B}{R_A + 2R_B} \tag{6.6}$$

Usually, duty cycle will be expressed as a percentage:

$$D\% = \frac{R_B}{R_A + 2R_B} \times 100\% \tag{6.7}$$

Note. If we try to obtain a square wave *i.e.*, $D = 50\%$ using this circuit, then excessive current will flow through the circuit because $R_A = 0$. This will spoil the entire circuitry. Some modifications need to be done in order to get a square wave using this circuit.

The typical applications of the astable operation of timer ICs include *FSK* generator, pulse position modulator, Schmitt trigger, free-running ramp generator, square wave generator, tone burst oscillator, and voltage-controlled frequency shifter.

In the case of astable operation, both states are quasi-stable *i.e.*, output will change from one state to another after some time without application of any external pulse.

6.4. APPLICATIONS

Linear Ramp Generator

The operation is monostable, but instead of the resistor R discussed in monostable operation, the constant current source is used.

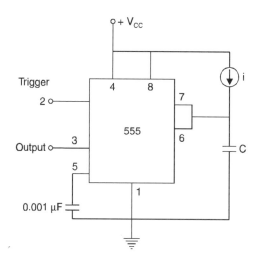

Fig. 6.7. Linear ramp generator.

Whenever the voltage at the trigger input is lower than $1/3\ V_{CC}$, the output at pin 3 attains a monostable state and capacitor C is charged by the constant current source. Hence, the charging is linear and the voltage at capacitor C is a linear ramp.

Time period T can be changed by changing capacitor C or the constant current source. This will cause the slope of the waveform to be changed.

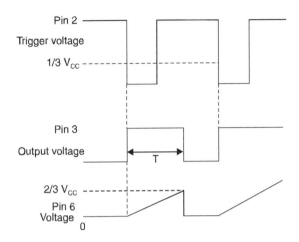

Fig. 6.8. Waveforms in a linear ramp generator.

Missing Pulse Detector

This also employs the monostable operation of IC 555. The trigger input is given external pulses, in which there is a missing pulse that has to be detected.

As long as the trigger input is receiving pulses, the output of IC 555 will have a monostable output state. If a pulse is not given to the trigger, then the circuit operates in its normal monostable operation.

The important thing is to set the time period of monostable operation to something larger than the time period of incoming pulses at the trigger input. If a new pulse is given before the completion of monostable operation, the output is forced to go to the monostable state with a new time period of oscillation. This is achieved by forcing the capacitor to discharge completely whenever there is an incoming pulse. The monostable operation will return to the stable output only after the capacitor charges to 2/3 V_{CC}. Whenever the incoming pulse is applied, the capacitor is completely discharged to 0 volts. Before the capacitor charges to 2/3 V_{CC}, the next pulse is applied so capacitor is discharged to 0 volts immediately. This will continue as long as the trigger has incoming pulses regularly. If any pulse is missing, then normal monostable operation will take place and the output will return to a stable state.

The practical circuit for this application is given in Figure 6.9.

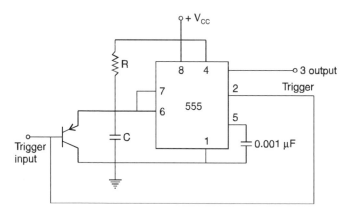

Fig. 6.9. Missing pulse detector.

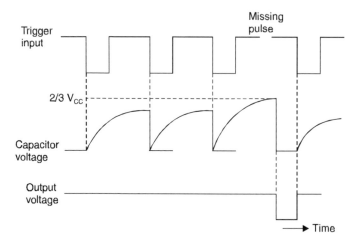

Fig. 6.10. Waveforms in a missing pulse detector.

Frequency Divider

A frequency divider circuit can be constructed as a mono-stable operation of IC 555 for which the trigger input is given by a square wave oscillator.

Note. If the trigger input is activated when the capacitor is charging *i.e.,* in a monostable state, nothing will happen to the monostable time period. In the example of the missing pulse detector, the trigger input affected the monostable operation by externally discharging the capacitor through the transistor. Here, no transistor is applied and nothing happens when a trigger pulse is applied in the middle of monostable operation.

By properly selecting T, the required frequency division can be achieved

$$f_0 = \frac{f_{in}}{K}$$

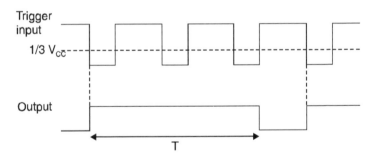

Fig. 6.11. Intermediate trigger input pulses not affecting monostable operation.

Rectangular Wave Generator

This is an astable application of timer ICs. For any general value of R_A and R_B, the simple circuit shown for astable operation is going to produce waveform having T_{High} and T_{Low} with different values. Hence, the simple circuit shown for astable operation will be useful as a rectangular wave generator.

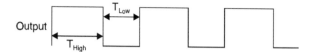

Fig. 6.12. Rectangular wave generator.

Schmitt Trigger

This is an astable operation of a timer IC.

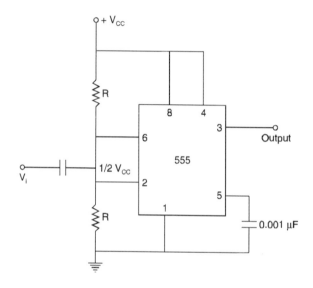

Fig. 6.13. Schmitt trigger circuit.

Both the lower comparator and upper comparator are biased with the external DC voltage $1/2\ V_{CC}$ through a potential divider built using the resistors.

If the input signal has a magnitude of $1/6\ V_{CC}$ (to be precise, a little more than $1/6\ V_{CC}$), then the voltage at pins 2 and 6 will be $2/3\ V_{CC}$. The upper comparator is triggered and *output = high*.

Whenever the input is less than $-1/6\ V_{CC}$ (*i.e.*, total voltage at pins 2 and 6 is less than $1/3\ V_{CC}$), then the lower comparator is triggered and *output = low*.

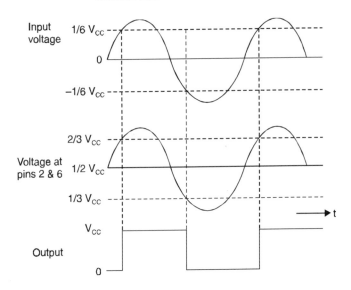

Fig. 6.14. Waveforms in Schmitt trigger circuit.

Whenever the input crosses 1/6 V_{CC}, the output voltage changes. Hence, the circuit acts as a Schmitt trigger.

7

DACs and ADCs

7.1. BASIC PRINCIPLES

Many measuring systems measure in analog form and all the physical quantities exist in analog form. Today, however, all the processing is done through computers which use digital technology. Hence, there is a need to convert analog data to digital data and vice versa.

This is why we need Digital to Analog Converters (DAC) and Analog to Digital Converters (ADC). As a group, they are Data Converters.

7.2. TECHNIQUES FOR DACs

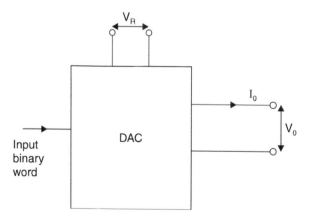

Fig. 7.1. Digital to analog convertor block diagram.

The DAC output can be either in the form of current I_0 or in the form of voltage V_0.

For this example, output in voltage is considered. The output voltage can be written:

$$V_0 = K \times V_{FS} (d_1 2^{-1} + d_2 2^{-2} + ... + d_n 2^{-n}) \qquad (7.1)$$

where V_0 = output voltage

K = multiplication factor, *i.e.*, scale factor

V_{FS} = output voltage for maximum input, *i.e.*, full-scale output voltage

$d_1, d_2 ... d_n$ = n-bit binary word

d_1 = most significant bit of binary word

d_n = least significant bit of binary word

The different techniques available for DACs are discussed below.

Weighted Resistor DAC

A weighted resistor DAC uses n switches for n different bits of binary word. It uses an operational amplifier in inverting amplifier mode and the resistors which are placed in different paths are multiplied by 2. This circuit is shown in Figure 7.2.

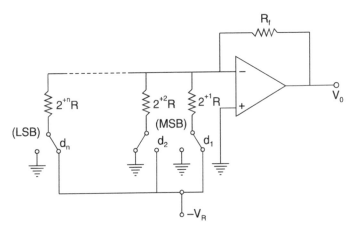

Fig. 7.2. Weighted resistor DAC.

$$V_0 = V_R \frac{R_f}{R} [d_1 2^{-1} + d_2 2^{-2} + ... + d_n 2^{-n}] \qquad (7.2)$$

The major disadvantages are that this requires a negative reference voltage and different values of resistors.

R-2R Ladder DAC

In an R-2R ladder DAC, only two values of resistor are used. The typical circuit is shown in Figure 7.3.

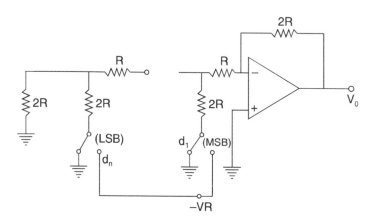

Fig. 7.3. R-2R ladder DAC.

The disadvantage is that if the input binary data is changed, the current flowing through the resistors changes and power dissipation across the resistors changes depending on the input binary word.

Inverted R-2R Ladder DAC

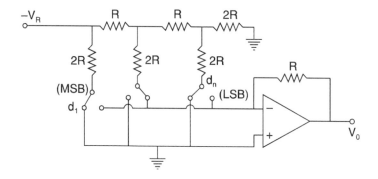

Fig. 7.4. Inverted R-2R ladder DAC.

An inverted R-2R ladder DAC is the same as an R-2R ladder DAC except that the current flow through the resistors doesn't depend on the input binary word and LSB and MSB are switched.

Multiplying DACs (MDAC)

This DAC uses a varying reference voltage V_R. Some typical examples of DACs are discussed below.

MC 1408 L is an 8-bit DAC with the output in the form of current.

SE/NE 5018 is an 8-bit DAC, with the output in the form of voltage.

$$
\text{DAC} \quad \left.\begin{array}{l} 0800/0801/0802 \\ 0830/0831/10832 \end{array}\right\} \quad \text{8-bit DAC}
$$

DAC   1200/1201/1202/1208/1209/1210—12-bit DAC

7.3. TECHNIQUES FOR ADCs

The function of an ADC is exactly opposite to that of a DAC. The output will have n bits, corresponding to n bits of binary output.

In most cases, a DAC will be included in ADC circuits to compare the generated output.

Flash ADC

This contains a 2^n to n encoder, a 2^n comparator, and their reference voltages.

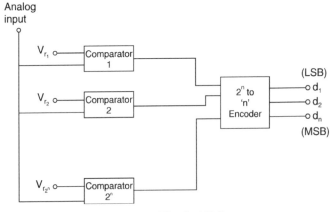

Fig. 7.5. Flash ADC.

Each comparator compares the reference voltage and the input voltage and produces an output corresponding to the voltages, which is encoded to produce the correct digital output. This has the advantage of being very fast, but the disadvantage of needing 2^n comparators and other accessories, which are costly.

Counter Type ADCs

A counter type ADC uses a counter that is forced to increase until the output is the exact digital equivalent of the given analog input. The circuit is shown in Figure 7.6.

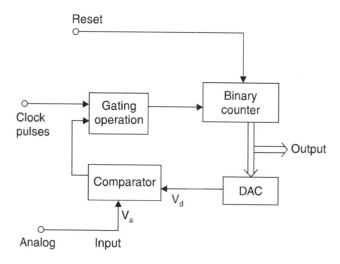

Fig. 7.6. Counter type ADC.

Initially, the counter is reset by applying a reset pulse. As long as the analog equivalent of the counter's output is lower than the analog input, the counter is increased.

Once the analog equivalent of counter's output V_d is greater than the analog input V_a, the counter is stopped and the output is displayed.

The disadvantage of this method is that if the given analog value is high, it will take a long time since the counter can only be increased by one each time.

Servo Tracking ADC

This is an improved version of the counter type ADC because the counter can count up or down. It should be noted that the counter can be increased or decreased, but only by one count.

The circuit is the same as that of the counter type, except the counter is replaced with an up/down counter.

Even this has the disadvantage that the tracking error, which is the difference between the actual input and the analog version of the output, is large if the input changes quickly.

Successive Approximation Counter

A successive approximation counter uses an efficient search method, reaching the final output in just n clock pulses, where n is the number of output bits.

First, the MSB is set to one and compared with the analog input. If the digital version has a greater magnitude, that bit is set to zero and the next bit is searched. If the digital version has a lower magnitude than the analog input, that bit is set to one and the next bit is searched.

The next bit is set to one and a comparison is done between the analog input and the digital output. This process continues until LSB is checked.

Integrating Type ADC

These devices produce an output value proportional to the analog input averaged over the integration period and do not require any sample and hold circuit.

Two techniques employed for this type are dual slope ADC and charge balancing ADC.

Typical ICs commercially available are discussed below:

ICL 7109 is a 12-bit dual slope ADC.

AD 7582 is a 12-bit DAC using successive approximation techniques.

AD 7520 and AD 7530 are 10-bit binary multiplying type ADCs.

AD 7521 and AD 7531 are 12-bit binary multiplying type ADCs.

ADC 0800, 0801, and 0802 are 8-bit ADCs.

8

Fundamentals of Digital Systems

8.1. INTRODUCTION

Digital Circuits. A digital signal is defined as a signal that has one of two possible states, either logic-0 or logic-1.

Digital circuits have digital signals for both input and output.

8.2. NUMBER SYSTEMS AND CONVERSION

A system in which there are only two output states is called a binary system. The value is either 0 or 1.

Conversion from Decimal System to Binary System

(*a*) **Full Number.** The given number is divided by 2 and the quotient and the remainder are noted. Then the quotient is divided by 2. This is done until the quotient becomes 0. Then by writing the remainders in order from the last remainder to the remainder from the first division, the binary equivalent of the given decimal is seen.

Let us consider the example of $(54)_{10}$

$$
\begin{array}{r|l}
2 & 54 \\
\hline
2 & 27 - 0 \\
\hline
2 & 13 - 1 \\
\hline
2 & 6 - 1 \\
\hline
2 & 3 - 0 \\
\hline
2 & 1 - 1 \\
\hline
 & 0 - 1 \\
\end{array}
\qquad (110\ 110)_2
$$

$\backslash$ $\qquad (54)_{10} = (110\ 110)_2.$

(*b*) **Fraction.** The given fraction is multiplied by 2, until the fraction becomes 0. The integers of each product are noted and are removed from the next stage of multiplication. Finally, the integers are written in order to get the binary equivalent.

If irrational fractions are to be converted into equivalent binary fractions, then the process is stopped, after noting that the fractions got in multiplication are repeated successively.

Let us consider the example of $(0.625)_{10}$.

$$\text{Integer}$$

$$
\begin{array}{ll}
0.625 \times 2 = 1.250 & 1 \\
0.250 \times 2 = 0.500 & 0 \\
0.500 \times 2 = 1.000 & 1
\end{array}
$$

Therefore, $(0.625)_{10} = (0.101)_2$

Let us consider another example of $(0.333)_{10}$.

$$\text{Integer}$$

$$
\begin{array}{ll}
0.333 \times 2 = 0.666 & 0 \\
0.666 \times 2 = 1.332 & 1 \\
0.332 \times 2 = 0.664 & 0 \\
0.664 \times 2 = 1.328 & 1 \\
0.328 \times 2 = 0.656 & 0 \\
0.656 \times 2 = 1.312 & 1 \\
0.312 \times 2 = 0.624 & 0 \\
0.624 \times 2 = 1.248 & 1 \\
0.248 \times 2 = 0.496 & 0 \\
0.496 \times 2 = 0.992 & 0 \\
0.992 \times 2 = 1.984 & 1
\end{array}
$$

Since the series is not converging, we can approximate:

$$(0.333)_{10} = (0.01010101001...)_2$$

(*c*) **Mixed Number.** When dealing with a fraction and an integer, the conversions are done separately and then they are added together.

Let us consider the example $(54.625)_{10}$

$$= (54)_{10} + (0.625)_{10}$$

These two numbers were used in the previous examples.

$(54)_{10} \quad = (110\ 110)_2$

$(0.625)_{10} \quad = (0.101)_2$

$\quad (54.625)_{10} = (110\ 110)_2 + (0.101)_2$

$\quad\quad\quad\quad = (110\ 110.101)_2$

Conversion from Binary to Decimal

Each digit of the binary number is assigned a positional of 2^n. For the first digit to the left of the decimal, $n = 0$. The digit to its left has $n = 1$ and so on. The first digit to the right of the decimal has $n = -1$. The next digit has $n = -2$ and so on. (See Example 1.)

Each digit's positional value is multiplied by its absolute value. The sum of those products is the decimal equivalent of the binary number.

Example 1.

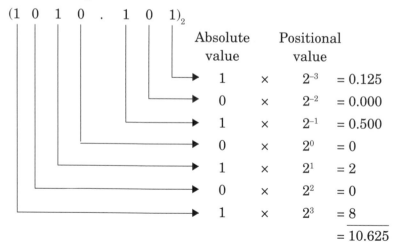

$(1 \quad 0 \quad 1 \quad 0 \quad . \quad 1 \quad 0 \quad 1)_2$

	Absolute value		Positional value	
	1	×	2^{-3}	= 0.125
	0	×	2^{-2}	= 0.000
	1	×	2^{-1}	= 0.500
	0	×	2^0	= 0
	1	×	2^1	= 2
	0	×	2^2	= 0
	1	×	2^3	= 8
				= 10.625

Therefore, $(1010.101)_2 = (10.625)_{10}$.

Other Number Systems

Hexadecimal System. Here the base is 16 and the symbols *A, B, C, D, E,* and *F* are assigned the values 10, 11, 12, 13, 14, and 15, respectively.

Octal System. Conversions from decimal to octal and from octal to decimal are the same as binary conversions, except the base is 8 instead of 2.

Conversion from Binary to Octal

By joining binary digits in groups of three starting from the decimal point and going left for integers and right for fractions and then putting the equivalent octal number for each binary combination, the equivalent octal number can be seen.

Example 2.

$(10101010 \, . \, 10101001)_2$

<u>10</u>	<u>101</u>	<u>010</u>	<u>101</u>	<u>010</u>	<u>01</u>
–	–	–	–	–	–
2	5	2	3	2	2

Therefore, $(10101010.10101001)_2 = (252.322)_8$.

Binary Coded Decimal System

The maximum value of a single number in the decimal system is 9, whose equivalent binary number is '1001'. To have uniformity, binary code digits are given in four digits, even though the decimal number may not need four digits in binary.

The table below shows the relationships between the systems discussed in this section.

Decimal No.	Binary Code	BCD	Octal	Hexa decimal
0	0	0000	0	0
1	1	0001	1	1
2	10	0010	2	2
3	11	0011	3	3

4	100	0100	4	4
5	101	0101	5	5
6	110	0110	6	6
7	111	0111	7	7
8	1000	1000	10	8
9	1001	1001	11	9
10	1010	0001 0000	12	A
11	1011	0001 0001	13	B
12	1100	0001 0010	14	C
13	1101	0001 0011	15	D
14	1110	0001 0100	16	E
15	1111	0001 0101	17	F
16	10000	0001 0110	20	10

8.3. POSITIVE LOGIC AND NEGATIVE LOGIC

If the voltage corresponding to logic-1 is higher than the voltage corresponding to logic-0, then it is said to have positive logic.

If the voltage corresponding to logic-0 is higher than the voltage corresponding to logic-1, then it is said to have negative logic.

8.4. SYNCHRONOUS AND ASYNCHRONOUS CIRCUITS

If all the components in a digital circuit are triggered simultaneously, then the circuit is synchronous. In other words, if the individual components of a digital circuit have clock input, then there will be a common path for clock input for all individual components in a synchronous circuit.

If there is no common clock pulse in the circuit, then it is asynchronous. In asynchronous circuits, all individual components are triggered at different times.

8.5. LOGIC GATES

Logic gates are the elements for performing logic functions. Examples of logic gates are discussed below.

OR Gate

No. of input terminals   = 2 or more

No. of output terminals   = 1

Symbol

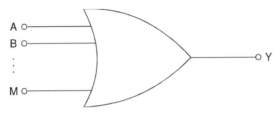

Fig. 8.1. Symbol of an OR gate.

Truth Table. A truth table is a table of all possible combinations of variables in the input, showing the relationship between the input and output. A truth table is written for a 2-input OR gate and a 3-input OR gate.

2-input OR Gate

A	B	Y
0	0	0
0	1	1
1	0	1
1	1	1

3-input OR Gate

A	B	C	Y
0	0	0	0
0	0	1	1
0	1	0	1
0	1	1	1
1	0	0	1
1	0	1	1
1	1	0	1
1	1	1	1

Equation

An OR gate performs **"Logical Addition"** and the output of an OR gate can be written as:

$$Y = A + B + \text{......} + M, \text{ for } M \text{ inputs.}$$

Note: Logical addition is different from binary addition.

Output Statement

Output of an OR gate = $\begin{cases} 1, \text{if any of the inputs} = 1 \\ 0, \text{if all the inputs} = 0 \end{cases}$

AND Gate

No. of input terminals = 2 or more

No. of output terminals = 1

Symbol

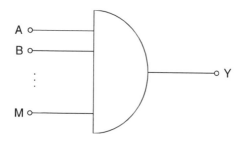

Fig. 8.2. Symbol of an AND gate.

Truth Table

2-input AND Gate

A	B	Y
0	0	0
0	1	0
1	0	0
1	1	1

3-input AND Gate

A	B	C	Y
0	0	0	0
0	0	1	0
0	1	0	0
0	1	1	0
1	0	0	0
1	0	1	0
1	1	0	0
1	1	1	1

Equation

An AND gate performs **"Logical Multiplication."** The equation for an AND gate can be represented as

$$Y = A \times B \dots M.$$

Output Statement

Output of an AND gate = $\begin{cases} 0, \text{if any of the inputs} = 0 \\ 1, \text{if all the inputs} = 1 \end{cases}$

NOT Gate

No. of input terminals   = 1

No. of output terminals   = 1

Symbol

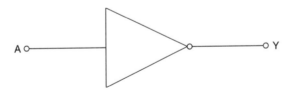

Fig. 8.3. Symbol of a NOT gate.

Truth Table

A	Y
0	1
1	0

Equation

A NOT gate performs **"Logical Negation"** or **"Logical Complementation."**

$$Y = A' = \overline{A}$$

Output Statement

Output of a NOT gate = $\begin{cases} 0, \text{if input} = 1 \\ 1, \text{if input} = 0 \end{cases}$

NOR Gate

No. of input terminals = 2 or more

No. of output terminals = 1

Symbol

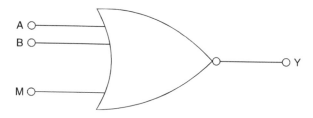

Fig. 8.4. Symbol of a NOR gate.

This is an OR gate followed by a NOT gate.

Truth Table

2-input NOR Gate

A	B	Y
0	0	1
0	1	0
1	0	0
1	1	0

3-input NOR Gate

A	B	C	Y
0	0	0	1
0	0	1	0
0	1	0	0
0	1	1	0
1	0	0	0
1	0	1	0
1	1	0	0
1	1	1	0

Equation

A NOR gate performs **"Logical Addition"** first, then **"Logical Negation."**

$$Y = (A + B + ... + M)^1$$

$$= \overline{(A + B + ... + M)}$$

Output Statement

$$\text{Output of a NOR gate} = \begin{cases} 1, \text{if all the inputs} = 0 \\ 0, \text{if any of the inputs} = 1 \end{cases}$$

NAND Gate

No. of output terminals = 2 or more

No. of input terminals = 1

Symbol

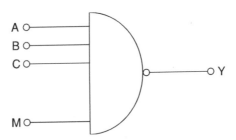

Fig. 8.5. Symbol of a NAND gate.

This is an AND gate followed by a NOT Gate.

Truth Table

2-input NAND Gate

A	B	Y
0	0	1
0	1	1
1	0	1
1	1	0

3-input NAND Gate

A	B	C	D
0	0	0	1
0	0	1	1
0	1	0	1
0	1	1	1
1	0	0	1
1	0	1	1
1	1	0	1
1	1	1	0

Equation

A NAND gate performs "**Logical Multiplication**" first, then "**Logical Negation**."

$$Y = (A \times B \dots M)^1$$

$$= \overline{(A \times B \dots M)}$$

Output Statement

Output of a NAND gate = $\begin{cases} 1, \text{ if any of the inputs} = 0 \\ 0, \text{ if all the inputs} = 1 \end{cases}$

Exclusive OR (Ex-OR) Gate

No. of input terminals   = 2

No. of output terminals = 1

Symbol

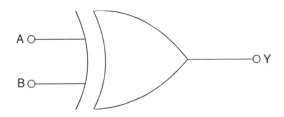

Fig. 8.6. Symbol of an Ex-OR gate.

Truth Table

A	B	Y
0	0	0
0	1	1
1	0	1
1	1	0

Equation

$$Y = A \oplus B$$

Output Statement

$$\text{Output} = \begin{cases} 1, \text{if there is a difference between the two inputs} \\ 0, \text{if there is no difference between the two inputs} \end{cases}$$

Note. NAND and NOR gates are known as "**Universal Gates**" because with the help of NAND gates or NOR gates, any given function can be realized.

8.6. FLIP FLOPS

Flip flops (or bitstable multivibrators) can maintain a binary state until directed by an input (clock) signal to switch states.

They can store either 0 or 1.

There are two output terminals Q and Q' (or $\overline{Q}$).

If $Q = 0$, then $Q' = 1$ and vice versa.

Clock Signals are binary control signals.

Note. If the clock signal is not activated, output will not change even if all the other inputs are changing.

The four types are RS Flip Flop, D Flip Flop, JK Flip Flop, and T Flip Flop.

Four Ways of Activating a Clock Signal in a Flip Flop

$clk = 1$ activation (clock high activation)

$clk = 0$ activation (clock low activation)

$clk = 1$ to 0 transition activation (trailing edge or negative edge activation or $^-$)

$clk = 0$ to 1 transition activation (leading edge or positive edge activation or -)

All flip flops can be activated, or triggered, in any one of these modes.

With four types of flip flop and four activation modes available, there are sixteen possible combinations.

Differences Between Logic Gates and Flip Flops

1. There are two outputs in a flip flop, but there is only one output in a logic gate.

2. A clock signal is present in a flip flop, but it is not needed for a logic gate.

Symbol of Flip Flops

The general symbol shows a rectangular box with the inputs on the left, the outputs on the right boxes, the clock input, and a corresponding symbol for the type of activation of the clock.

RS FLIP FLOP

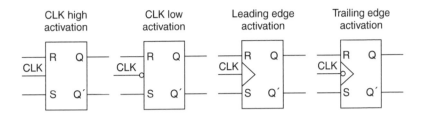

JK FLIP FLOP

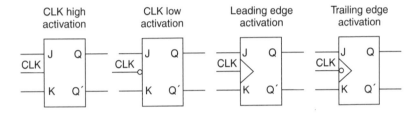

D FLIP FLOP

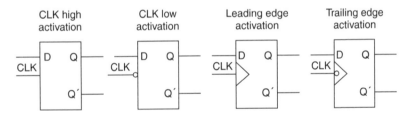

Fig. 8.7. Flip flops.

Clock Activation Circuit

Note. When activated, a clock signal has an amplitude equal to logic-1.

A CLK high activation circuit is constructed using AND gates only.

A CLK low activation circuit is constructed using AND gates and a NOT gate which will preceed the AND gates.

A leading edge activation circuit is constructed using AND gates and a differentiator.

A trailing edge activation circuit is constructed using AND gates, a differentiator, and a NOT gate between the differentiator and the AND gates.

Truth Table

RS Flip Flop

R	S	Q
0	0	No change
0	1	1
1	0	0
1	1	Not possible

D Flip Flop

D	Q
0	0
1	1

JK Flip Flop

J	K	Q
0	0	No change
0	1	1
1	0	0
1	1	Complement (Toggle)

T Flip Flop

T	Q
0	No change
1	Complement (Toggle)

8.7. SHIFT REGISTERS

A register is a group of flip flops that can be used to store a given binary number. A shift register is a circuit of connected flip flops. After each clock pulse, the data from one flip flop is moved, or shifted, serially one bit to the left or to the right, depending on the connection between flip flops.

Classification

Depending on the direction of shift of the data, the shift registers can be classified as

    (i) shift right shift register,

    (ii) shift left shift register, or

    (iii) bidirectional shift register.

In the case of shift right shift register, the direction of data shift is towards the right hand side. Data enters the left most point of the register and leaves through the right most point after the application of sufficient clock pulses.

In the case of shift left shift register, the direction of the data shift is towards the left hand side. Data enters the right most point and leaves through the left most point after the application of sufficient clock pulses.

Bidirectional shift registers can function as a shift left shift register or as a shift right shift register.

The types of input and output further classify shift registers. They are discussed below.

Serial In Serial Out Shift Register (SISO S.R.)

The input and the output for this device are serial in nature.

Let us consider the diagram in Figure 8.8. The values of Q_1, Q_2, Q_3, and Q_4 are 0, 0, 1, and 0, respectively. The value of the input at serial input pin $D_1 = 1$. Present output $= Q_4 = 0$, which is available at the serial output pin. Before application of the first pulse, $D_2 = Q_1$, $D_3 = Q_2$, and $D_4 = Q_3$.

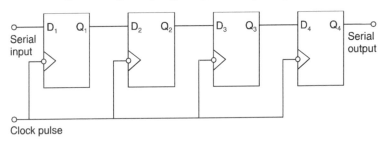

Fig. 8.8. Serial in serial out shift register.

After the application of the first clock pulse, all values shift one place to the right. This makes $Q_1 = D_2 = 1$, $Q_2 = D_3 = 0$, $Q_3 = D_4 = 0$, and serial output $= Q_4 = 1$.

These details are more easily seen in the form of a table:

Clock Pulse No.	Serial Input	Q_1	Q_2	Q_3	Q_4 (= Serial output)
0	1	0	0	1	0
1	0	1	0	0	1
2	0	0	1	0	0
3	1	0	0	1	0
4	1	1	0	0	1
5	0	1	1	0	0

Thus the data placed in serial input is moved serially, one bit at a time, inside the shift register.

The same can be shown in Fig. 8.9.

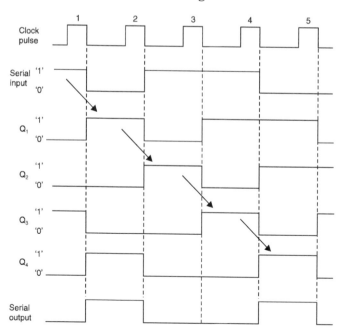

Fig. 8.9. Visual representation of SISO table.

Serial In Parallel Out Shift Register (SIPO S.R.)

The data input is in serial form and the output is in parallel form.

The operation of SIPO S.R. is the same as that of the SISO S.R. except that all the intermediate stage outputs, Q_1, Q_2, and Q_3, are also taken as output.

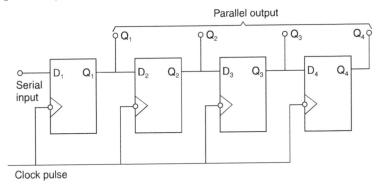

Fig. 8.10. Serial in parallel out shift register (SIPO S.R.).

Operation is the same as the SISO shift register.

Parallel In Serial Out Shift Register (PISO S.R.)

The input is given in parallel form and the output is taken out serially one bit at a time from the device.

The basic circuit is shown in Figure 8.11. It can be seen that this violates the basic rule of shift register, *i.e.*, the output of one stage is the input of the next stage. To make this a shift register, a serial connection must be made. This device is usually

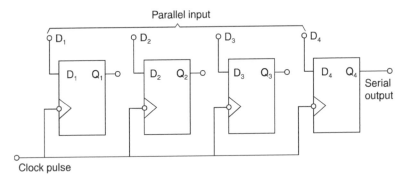

Fig. 8.11. Principle of parallel in serial out shift register.

used to give the parallel data input once and then it is used as a serial in serial out shift register.

A practical PISO S.R. can be drawn as shown in Figure 8.12. If the serial shift/parallel load is **high**, gates G_4, G_5, and G_6 are enabled as well as G_7, G_8, and G_9. Thereby **serial shift operation** is done.

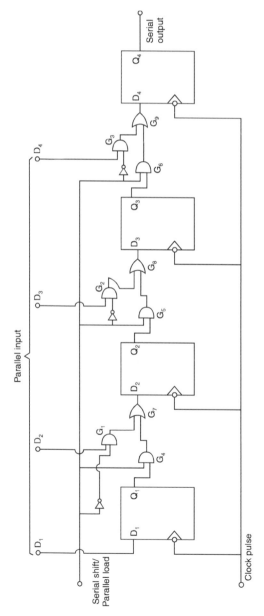

Fig. 8.12. Parallel in serial out shift register.

If the serial shift/parallel load is **low**, gates G_1, G_2, and G_3 are enabled as well as G_7, G_8, and G_9. Thereby **parallel load operation** is done.

We will assume an example where we will load the input data 1101 for D_1, D_2, D_3, and D_4 and the data are shifted serially later.

First we will load the data parallel, by making the serial shift/parallel load line input low. $D_1 = 1, D_2 = 1, D_3 = 0$, and $D_4 = 1$ are set. After application of the first clock pulse, $Q_1 = 1, Q_2 = 1$, $Q_3 = 0$, and $Q_4 = 1$.

Now serial shift/parallel load line input is switched to high, making this function like a serial in serial out shift register from this point forward.

Clock pulse	Serial shift / Parallel load	Data in D_1	Q_1	Q_2	Q_3	Q_4	Serial out (Q_4)
0	0	1	—	—	—	—	—
1	1	0	1	1	0	1	1
2	1	1	0	1	1	0	1
3	1	0	1	0	1	1	0
4	1	1	0	1	0	1	1

Parallel In Parallel Out Shift Register (PIPO S.R.)

The input and the output are parallel in form.

The operation of this circuit is the same as the PISO S.R. except that intermediate stage outputs are taken separately as final output.

The circuit for parallel in parallel out shift register is shown in Figure 8.13.

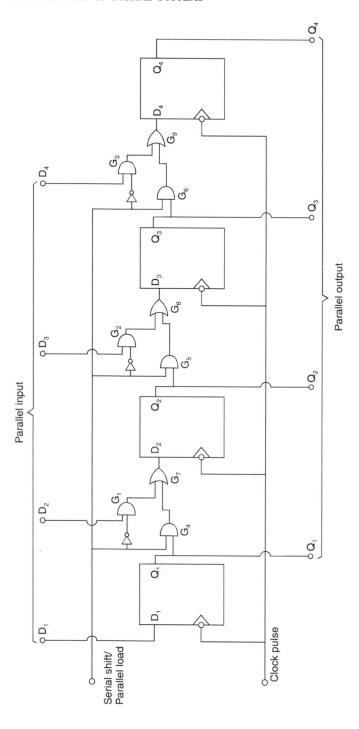

Fig. 8.13. Parallel in parallel out shift register.

8.8. COUNTERS

A sequential circuit that goes through a prescribed sequence of states upon the application of input pulses is called a counter.

It is used for counting the number of pulses applied to the device.

Classification

Counters are classified as either synchronous or asynchronous.

A counter that follows a binary sequence is called a binary counter.

Synchronous Counters

All the flip flops present in a synchronous counter are triggered simultaneously.

An example of a 3-bit synchronous binary counter is shown in Figure 8.14.

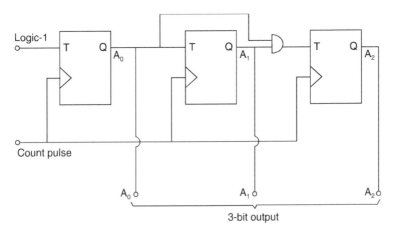

Fig. 8.14. 3-bit synchronous counter.

Before the first pulse, the states of A_2, A_1, and A_0 are 0, 0, and 0.

After receiving the first count pulse, A_0 becomes complemented, i.e., $A_0 = 1$, and A_1 and A_2 remain at 0.

Therefore, output = 001, which is the binary equivalent of 1, the number of count pulses applied so far.

After receiving the second count pulse, A_0 becomes complemented. This complements A_1. Therefore, $A_0 = 0$, $A_1 = 1$, and $A_2 = 0$.

Therefore, output = 010, which is the binary equivalent of 2, the number of count pulses applied so far.

After the third count pulse, A_0 is complemented. Therefore, $A_0 = 1$, A_1 = its previous value = 1, A_2 = its previous value = 0.

Therefore, output = 011.

After the fourth count pulse, A_0 is complemented. Therefore $A_0 = 0$, $A_1 = 0$, and $A_2 = 1$.

Therefore, output = 100.

Similarly, after the fifth count pulse, output = 101.

After the sixth count pulse, output = 110.

After the seventh count pulse, output = 111.

After the eighth count pulse, A_0 is complemented, which complements A_1, which complements A_2.

Therefore, output = 000.

The same information can be drawn as a waveform.

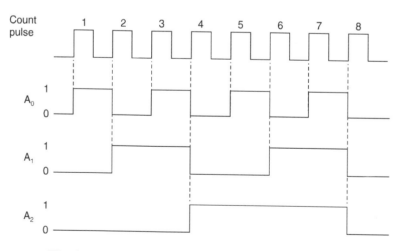

Fig. 8.15. Waveforms in a synchronous bit counter.

The same information can be written as a table.

Count pulse	A_2	A_1	A_0
0	0	0	0
1	0	0	1
2	0	1	0
3	0	1	1
4	1	0	0
5	1	0	1
6	1	1	0
7	1	1	1
8	0	0	0

Asynchronous Counters

There is no common path for the count signal to all the flip flops. (See Figure 8.16.)

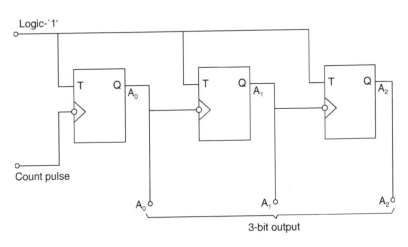

Fig. 8.16. 3-bit asynchronous counter.

Consider an example of a 3-bit asynchronous binary counter and look at Figure 8.17 and the accompanying table.

The table can be written as

Count Pulse	A_2	A_1	A_0
0	0	0	0
1	0	0	1
2	0	1	0
3	0	1	1
4	1	0	0
5	1	0	1
6	1	1	0
7	1	1	1

The waveforms can be drawn as shown in Figure 8.17.

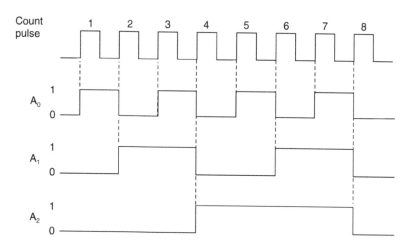

Fig. 8.17. Waveforms in 3-bit asynchronous counter.

Other examples of counters are Johnson counter, Binary up-down counter, Ripple counter, BCD counter, and counter with non-binary sequence.

8.9. EXPRESSION WRITING AND SIMPLIFICATION

Symbols for all logic gates have already been given. Simplification of expression can be done by using a technique called the Karnaugh map or K-map. This technique uses simplification using diagramatic reduction.

8.10. IMPLEMENTATION OF A GIVEN FUNCTION

It is easier to implement a given function backward rather than forward, *i.e.* from the last operation to the first.

Example 1

$$F = xyz + xyz' + x'yz$$

This function is the sum of 3 components. A sum of 3 components requires a 3-input OR gate. In this case, the inputs are xyz, xyz', and $x'yz$.

The next step is to implement the individual components of the OR gate inputs separately. This requires three 3-input AND gates (since a product requires an AND gate)—one for each input of the OR gate.

For the first term, the inputs are $x, y,$ and z. For the second term, the inputs are $x, y,$ and z'. For the third term, the inputs are $x', y,$ and z.

As a final step, all the x inputs are joined together and shown as a single input x. Wherever x' is needed, a NOT gate is used since it will give complemented output.

The same will be true for all the y inputs and z inputs.

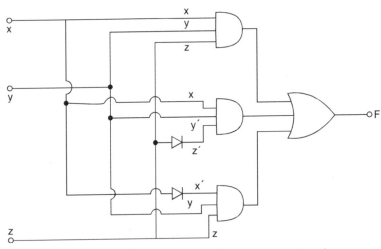

Fig. 8.18. Implementation of $F = xyz + xyz' + x'yz$.

Example 2

$$F = (x + y)(x + z)(x' + z)$$

Step 1. AND gate having 3 inputs—$x + y$, $x + z$, and $x' + z$

Step 2.

Term 1 — OR gate with inputs x and y

Term 2 — OR gate with inputs x and z

Term 3 — OR gate with inputs x' and z

Step 3. Combining all x, y, and z inputs together.

Step 4. NOT gate for all x', y' and z' inputs.

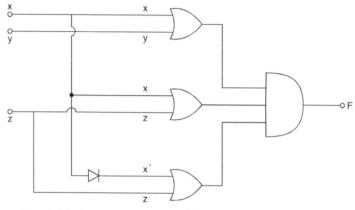

Fig. 8.19. Implementation of $F = (x + y)(x + z)(x' + z)$.

Digital Circuits—I

9.1. ADDER

An adder is a digital circuit that performs addition of digital inputs.

A simple addition of two binary bits can cause any one of four possible elementary operations, namely $0 + 0 = 0$, $0 + 1 = 1$, $1 + 0 = 1$, and $1 + 1 = 10$. The first three operations produce a sum whose length is one bit. The fourth operation produces a sum whose length is 1 bit and a carry whose length is 1 bit. A higher order bit which is a result of addition is called a *carry*.

An adder circuit that performs the addition of two bits is called a *half-adder*. An adder circuit that performs the addition of three bits is called a *full-adder*.

Half-Adder

Inputs		Output	
x	y	*Carry*	*Sum*
0	0	0	0
0	1	0	1
1	0	0	1
1	1	1	0

Writing Boolean expressions for carry and sum outputs, after simplification using the K-map technique, is simple. The final expression for carry and sum may be derived as follows:

Sum (S) $= x'y + xy'$

Carry (C) $= xy$.

Implementation for the two outputs are done separately. As a final step, the inputs x and y are combined.

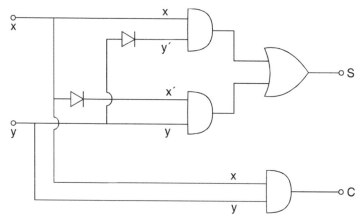

Fig. 9.1. Half-adder circuit.

Note. The simplification of a Boolean expression may be done in different ways and there can be other ways of implementing a half-adder circuit.

Full-Adder

Out of the three binary inputs added in a full-adder circuit, two inputs are used as external inputs and the third is used as a carry in the previous stage.

x	y	C_{in} or 'Z' (carry in)	C_{out} (carry out)	Sum
0	0	0	0	0
0	0	1	0	1
0	1	0	0	1
0	1	1	1	0
1	0	0	0	1
1	0	1	1	0
1	1	0	1	0
1	1	1	1	1

After writing and simplifying the Boolean expression for C_{out} and S, the equations can be written as follows:

$$C_{out} = xy + xz + yz$$
$$S = x'y'z + x'yz' + xy'z' + xyz$$

Implementation of these two expressions may be done as was done for the half-adder. The circuit may be drawn as shown in Figure 9.2.

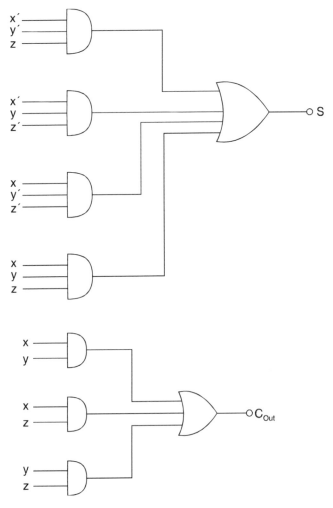

Fig. 9.2. Full-adder circuit.

9.2. CARRY LOOK-AHEAD ADDER

The full-adder circuit equations can also be written

$$S = P \oplus Z$$

and $$C_{out} = G + PZ \quad \text{where } P = x \oplus y \text{ and } G = x \times y.$$

G is called a **carry generate** and it produces an output carry when both inputs x and y are equal to one regardless of the input carry z.

P is called a **carry propagate**, because it is the term associated with the propagation of the carry from input level Z to output level C_{out}.

Realizing those equations, the circuit for a look-ahead adder can be drawn as shown in Figure 9.3.

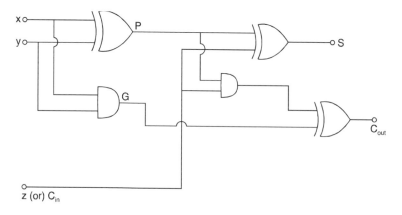

Fig. 9.3. Look-ahead adder.

9.3. PARITY CHECKER AND GENERATOR

A parity bit is added to the message bits during the transmission of binary information for the purpose of detecting errors.

In practical transmission systems, there is a possibility of errors in the transmission channel. To reduce the errors in transmission, a parity bit is added.

If the total number of 1's in the message, including the parity bit, is even, then the system is called an **even parity**

system. If the total number of 1's in the message including the parity bit is odd, then the system is called an **odd parity system**.

A circuit that generates the parity bit in the transmission is called a **parity generator**.

A circuit that checks the parity bit on the receiver side is called a **parity checker**.

Let us assume that a message block is four bits. With the parity bit, five bits should be transmitted through the transmission channel.

Let the message bits be designated with the symbols A, B, C, and D and the parity bit with the symbol P.

Parity Generator

The required truth table is written first. All the input combinations are assumed and the corresponding parity bit is calculated. The parity assumed here is even parity.

A	B	C	D	No. of ones excluding parity	Parity bit to be added to make no. of 1's to even
0	0	0	0	0	0
0	0	0	1	1	1
0	0	1	0	1	1
0	0	1	1	2	0
0	1	0	0	1	1
0	1	0	1	2	0
0	1	1	0	2	0
0	1	1	1	3	1
1	0	0	0	1	1
1	0	0	1	2	0
1	0	1	0	2	0
1	0	1	1	3	1
1	1	0	0	2	0
1	1	0	1	3	1
1	1	1	0	3	1
1	1	1	1	4	0

The simplest technique to produce a parity bit is to use an Ex-OR gate.

$$P = A \oplus B \oplus C \oplus D$$

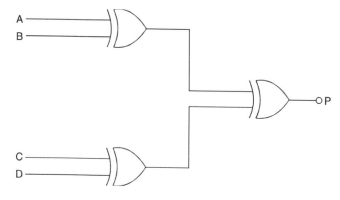

Fig. 9.4. Parity generator.

Parity Checker

This is placed on the receiver side. After receiving all five bits (four message bits + one parity bit), transmission errors will be detected using the equation

$$\text{Error } E = A \oplus B \oplus C \oplus D \oplus P$$

First, we will assume no transmission error and we will check the truth table.

If there are no transmission errors, the output of the parity checker is zero.

We will assume that the message bits are $A = 0$, $B = 1$, $C = 1$, $D = 1$, and the parity bit (sent) = 1. If there is a transmission error while sending message bit $D = 1$, then that bit will be received as $D = 0$.

∴   The bits received will be $A = 0$, $B = 1$, $C = 1$, $D = 0$ and $P = 1$. For this case, error can be calculated from the equation

$$E = A \oplus B \oplus C \oplus D \oplus P$$

or $$E = 0 \oplus 1 \oplus 1 \oplus 0 \oplus 1 = 1$$

∴   We can conclude that an error has occurred while transmitting the data.

A	B	C	D	Parity 'P'	Error 'E'
0	0	0	0	0	0
0	0	0	1	1	0
0	0	1	0	1	0
0	0	1	1	0	0
0	1	0	0	1	0
0	1	0	1	0	0
0	1	1	0	0	0
0	1	1	1	1	0
1	0	0	0	1	0
1	0	0	1	0	0
1	0	1	0	0	0
1	0	1	1	1	0
1	1	0	0	0	0
1	1	0	1	1	0
1	1	1	0	1	0
1	1	1	1	0	0

The circuit to realize this can be drawn as shown in Figure 9.5.

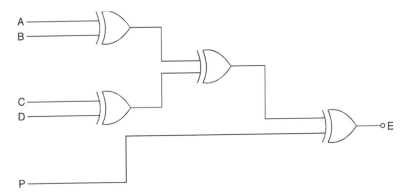

Fig. 9.5. Parity checker.

Conclusion. With this tactic, we can note whether an error has occurred while transmitting the data. This circuit can only detect an odd number of errors. It will not detect an even number of errors occurring in the transmission.

9.4. READ ONLY MEMORY (ROM)

Read only memory is a device in which permanent binary information is stored. Once written, the data stored cannot be rewritten, only read.

The data stored in the ROM is not volatile *i.e.,* is not affected when the device is switched ON and OFF.

The ROM has n input lines and m output lines. Each bit combination of the input variables is called an **address** and each bit combination from the output lines is called a **word**.

In other words, the ROM is capable of storing 2^n words of n bit length. By selecting any one of these 2^n words, a particular word corresponding to that address can be read.

Fig. 9.6. ROM block diagram.

We will start with an example of 16×8 ROM or $2^4 \times 8$ ROM. By selecting any one of these 16 addresses and varying the 4 input variables, we can read the data stored in a particular address. The number of bits per output word in this example is 8. This is shown in Figure 9.7.

The part shown inside the dotted lines is called a **decoder.**

Short-circuiting a fuse will store the bit '1' and open-circuiting a fuse will store the bit '0'.

By selectively short-circuiting desired fuses and open-circuiting other fuses, the corresponding data can be stored in ROM.

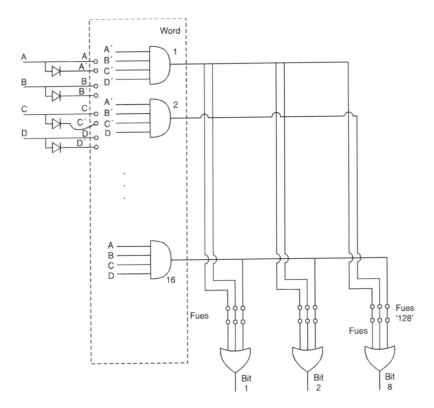

Fig. 9.7. Logic construction of 16 × 8 ROM.

9.5. PROGRAMMABLE ROM (PROM)

The user can write his own words with this type of ROM. The manufacturer will short-circuit all the fuses. The user can apply specific current pulses to the fuses that should be open-circuited. Care should be taken while applying current pulses because an open-circuited fuse cannot be short-circuited. This process is called **programming**.

Once programmed the word stored cannot be changed.

9.6. ERASABLE PROM (EPROM)

This type of PROM can be reset by exposing the device to ultraviolet rays. This process is called **erasing**. EPROM can be programmed and erased many times for different applications.

9.7. ELECTRICALLY ERASABLE PROM (EEPROM)

EEPROM functions the same as EPROM, but it uses electrical signals instead of ultraviolet rays.

9.8. PROGRAMMABLE LOGIC ARRAY (PLA)

A combinational circuit may occasionally have **don't care conditions**, when implemented with a ROM. A don't care condition is an address input that will never occur.

When the number of don't care conditions is excessive, it is a waste of equipment. It is more economical to get a programmable logic array (PLA).

Since all of the input combinations need not be used, fuses can be placed in the input side. Using this process, the number of gates in the system can be drastically reduced.

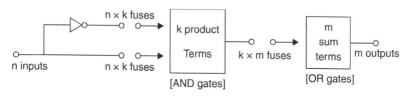

Fig. 9.8. PLA block diagram.

Instead of 2^n AND gates, we have reduced the AND gates on the input side to k, which is lower than 2^n.

Similarly, on the output side, instead of having $2^n \times m$ fuses, the number is reduced to $k \times m$ fuses.

Some PLAs can be programmed by the user. Those devices are called **field programmable logic arrays**, because programming is done in the field.

9.9. PROGRAMMING ARRAY LOGIC (PAL)

A PAL has a **fixed OR array** on the output side and a **programmable AND array** on the input side.

A PAL has less flexibility than a PLA because it does not allow the user to program the OR array.

A conventional case of an AND gate with 3 input fuses is shown in Figure 9.9.

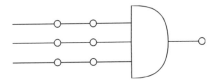

Fig. 9.9. Symbol of 3-input AND gate.

The same configuration in the case of a PAL is shown in Figure 9.10. The intersecting lines represent the fusing.

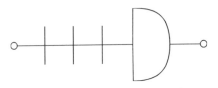

Fig. 9.10. Alternate symbol of 3-input AND gate.

A cross mark shows a particular contact.

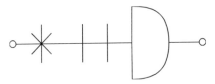

Fig. 9.11. Cross showing a fuse.

In Figure 9.11, fuse-1 is short-circuited and the other fuses are open-circuited.

In the following example, we will consider a 3-input, 3-output PAL.

Each output function will come from an OR gate at output level. The inputs for the OR gates come from outputs of AND gates, the inputs of which can be programmed.

Since there are three input terminals, six input combinations are possible. Those versions are connected to AND gates.

The crosses shown in Figure 9.12 indicate that the particular fuse has been short-circuited.

$$F_1 = BC + A'C'$$
$$F_2 = A'C + BC$$
$$F_3 = A'C + AB'C'$$

Note. The number of inputs to the OR gate has to be selected based on the application.

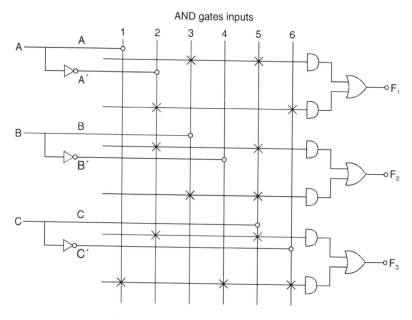

Fig. 9.12. A typical PAL circuit.

10

Digital Circuits—II

Memory can be classified into two types: read only memory and random access memory. Read only memory was discussed in chapter 9. Random Access Memory (RAM), or read/write memory, can be read and written at any time. There are no restrictions for writing the data into RAM. The two types of RAM are described below.

10.1. STATIC RAM (SRAM)

SRAM consists mostly of flip flops that store the binary information. The stored information remains valid as long as power is applied to the unit.

When compared with dynamic RAM (see section 10.2), SRAM is easier to use. It has shorter read and write cycles, can be constructed both in TTL and MOSFET technologies, and uses conventional storage devices like flip flops, whereas dynamic RAM requires charging or discharging capacitances.

10.2. DYNAMIC RAM (DRAM)

This type of RAM uses unconventional memory storage devices. Logic levels are preserved as a charge, or an absence thereof, on capacitances. The primary advantage of this technology is that more data can be stored in a small area. Another advantage is that it consumes less power than SRAM. The drawback to dynamic RAM is that the circuitry weakens over time and will lose its charge. Therefore, it must be refreshed periodically. Dynamic RAM is available only in MOSFET circuits.

10.3. CHARGE COUPLED DEVICE (CCD)

A CCD is a type of dynamic memory in which packets of charge are continuously transferred from one MOS device to another. CCD memory is inherently serial. In practice, memories are constructed in the form of shift registers, where each shift register is a line of CCDs. By controlling the timing of the clock signals applied to the shift registers, data can be accessed one bit at a time from a single register or several bits at a time from multiple registers.

The principal advantage of CCD memory is that it has a simple cell structure, making it possible to construct large capacity memories at low cost. But it also has the problem associated with dynamic RAM of needing the circuitry to be refreshed periodically.

10.4. MOORE AND MEALEY MACHINES

The output of Mealey Model Machines depends on both the present input and the present outputs. A shift resister is an example of a Mealey Machine.

The output of a Moore Model Machine is dependent solely upon the present output; no input is required. A counter is an example of a Moore Machine.

10.5. PSEUDO RANDOM BINARY SEQUENCE (PRBS) GENERATOR

A PRBS generator is a shift register with selected intermediate stage outputs that are given as input to a combinational logic device. The output from that device is used as feedback, or input, for the shift register.

The maximum length of the PRBS waveform is $2^N - 1$ bits, where N is the number of stages in the shift register. It can be obtained by choosing the proper tappings in the shift register. The frequency of the PRBS waveform is the same as the clock frequency of the shift register.

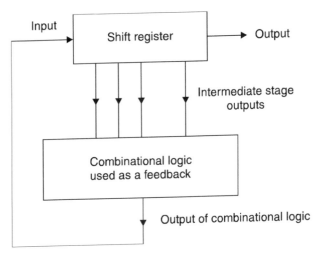

Fig. 10.1. Principle of a PRBS generator.

We will take an example.

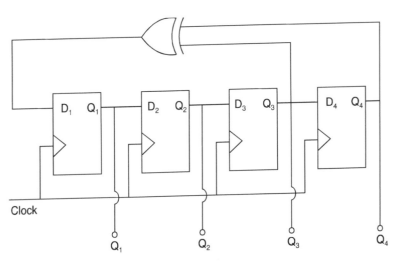

Fig. 10.2. A typical PRBS generator.

The truth table for this example is:

Q_1	Q_2	Q_3	Q_4	Output of Combinational logic
1	1	1	1	0
0	1	1	1	0
0	0	1	1	0
0	0	0	1	1
1	0	0	0	0
0	1	0	0	0
0	0	1	0	1
1	0	0	1	1
1	1	0	0	0
0	1	1	0	1
1	0	1	1	0
0	1	0	1	1
1	0	1	0	1
1	1	0	1	1
1	1	1	0	1
1	1	1	1	0 → same as first state

10.6. ACCUMULATOR

When multiplying or adding two data, there is a need for three registers, two for input data and one for storing output data. An accumulator shrinks this device to only two.

An accumulator is a special register that can process input data and then store the output after the operation is completed.

10.7. LIQUID CRYSTAL DISPLAY (LCD) DRIVERS

LCDs, along with LEDs (light emitting diode) are the most popular display devices.

LCDs require less power than LEDs, but they provide less illumination than LEDs. LCDs are used in calculators and other devices where power is the main constraint.

An LCD driver is a specialized IC which converts the given data into a format suitable for illumination through a 7-segment display. (See section 10.10.)

10.8. FREQUENCY COUNTERS

Frequency counters are specialized counters that count pulses which will be given as a signal.

By dividing the number of pulses in a given period by the time, the frequency of the pulses can be calculated.

Signals that are not in pulse type, will be converted into a pulse type waveform first and then the frequency can be measured.

10.9. 7-SEGMENT DISPLAY FOR FREQUENCY COUNTER

By arranging seven segments on an LED or LCD, all the numerals and some alphabets can be displayed. This 7-segment display is common for all sort of displays. Examples of 7-segment displays include digital watches and calculators.

The basic arrangement for a 7-segment display is shown in Figure 10.3.

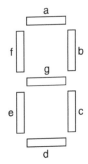

Fig. 10.3. 7-segment display.

The numerals displayed by the 7-segment display are shown in Figure 10.4.

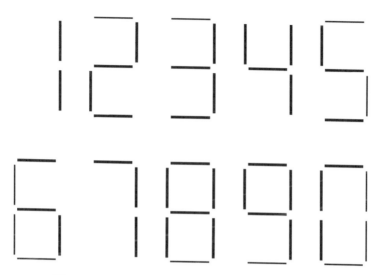

Fig. 10.4. Numbers using a 7-segment display.

The 7-segment display is classified as a common anode display or a common cathode display, depending on the type of connection.

In a common anode display, all the anodes are connected to the positive power supply. By giving a low voltage to the required segment, that segment will glow.

In a common cathode display, all the cathodes are connected to the negative power supply. By giving a high voltage to a particular segment, that segment will glow.

10.10. MICROPROCESSOR

A microprocessor is a multi-purpose, programmable logic device that reads binary instructions from a storage device called memory, accepts binary data as input and processes data according to those instructions and provides the result as output.

A microprocessor is the processing unit of a computer. The memory and input/output (I/O) devices are external and the microprocessor interacts with them. Typical examples of microprocessors are 8085, Z80, 8008, 8080, MC 6800, and HD 64180.

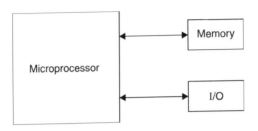

Fig. 10.5. Diagram of a microprocessor and components.

Microprocessors are classified based on their handling capability. For example, a 4-bit microprocessor can handle only four bits at a time.

10.11. MICROCONTROLLER

A microcontroller, or microcomputer, is a single IC chip containing a processor, timers, and memory. The typical applications of microcontrollers are industrial instrumentation, video games, word processing, and small-business applications.

A typical microcontroller has a microprocessor, 64 bytes of read/write memory, 1K ROM, and several signal lines to connect I/O's. Some examples are Z8, MCS 51 and 96 series, and 68 HC 11.

11

Logic Families

11.1. INTRODUCTION TO LOGIC FAMILIES

Logic gates are grouped into "families" and "systems" based on similarities in manufacturing process. Examples of logic families are Diode Logic (DL), Resistor Transistor Logic (RTL), Resistor Capacitor Transistor Logic (RCTL), Diode Transistor Logic (DTL), Transistor Transistor Logic (TTL), Emitter Coupled Transistor Logic (ECL), and Complementary Metal Oxide Semiconductor Logic (CMOS).

11.2. IMPORTANT CHARACTERISTICS OF LOGIC FAMILIES

Fan-In

The maximum number of inputs that can be applied to a logic gate is known as fan-in.

A 4-input NOR gate can have up to four input lines. So, a 4-input NOR gate has a fan-in of 4.

Fan-Out

Sometimes the output from one gate will be used as input for the next stage. The fan-out of a logic gate is the maximum number of gates that a particular logic gate can drive.

If a typical NOR gate has a fan-out of 8, then it can drive eight gates from its output.

Loading Factors

A gate's fan-in is called its input loading capability and the fan-out is its output loading capability. Together they are called loading factors.

Transfer Characteristics

The transfer characteristics of a logic gate are represented by a curve relating the output voltage to the input voltage. The curve is plotted on a graph where the input voltage is on the x-axis and the output voltage is on the y-axis.

So far, we have considered logic-1 and logic-0 as having a constant voltage level. In actuality, they do not have a single precise voltage level. They each represent a band of voltage and there is a band of voltage between those two voltage levels.

V_{OH}—The nominal, or minimum, logic-1 state output voltage.

V_{OL}—The nominal, or maximum, logic-0 state output voltage.

V_{iL} — The nominal, or maximum, input voltage required for logic-0 input.

V_{iH} — The nominal, or minimum, logic-1 input voltage.

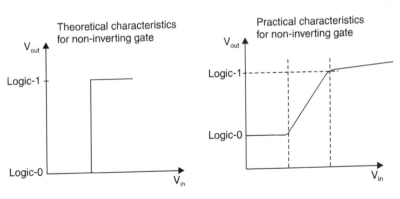

Fig. 11.1. Transfer characteristics of logic gates.

These parameters will vary, whenever there is a change in the supply voltage, temperature, loading variation, etc. These variations are called spreads in transfer characteristics.

Logic Swing

The difference between two output voltages is referred to as the logic swing of the circuit.

Logic swing = $V_{OH} - V_{OL}$

Noise Margin

Noise may be created at any point in the circuit. There can be internal noises, noises due to improper terminations, and others.

The maximum value of noise signals, which can enter the logic gate without affecting the performance of the logic gate, is known as noise margin.

For logic gates, there are two DC noise margins. The one for high input states is called logic-1 state noise margin and the one for low input states is called logic-0 state noise margin.

High state noise margin is defined as the difference between the logic-1 state output voltage, for a full fan-out load, and the logic-1 state input voltage.

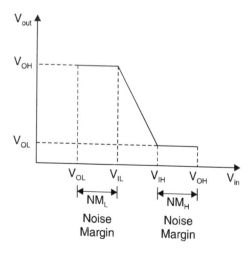

Fig. 11.2. Noise margin.

Low state noise margin is defined as the difference between logic-0 state input voltage and the logic-0 state output voltage.

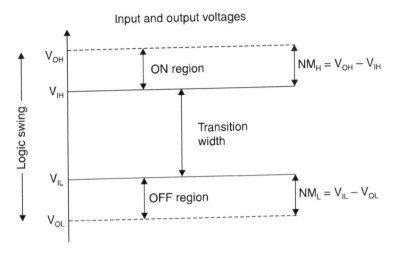

Fig. 11.3. Noise margin.

Noise Immunity

The noise immunity of a logic gate is the voltage which, if applied to the input, will cause the output to change the output state for a given supply voltage V_{CC}.

Propagation Delay

The time interval between a change in input state and the resulting change in output state is the propagation delay.

If there are many gates in a digital circuit, then the total propagation time delay in the circuit is the algebraic sum of the propagation delays of individual gates in the series.

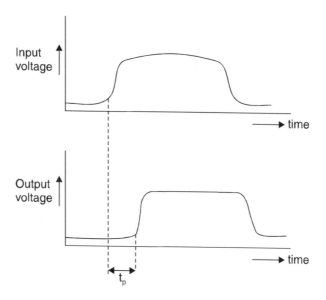

Fig. 11.4. Propagation delay.

Power Supply Specifications

The important power supply specifications are nominal voltage, regulation, output ripple, and load requirement.

Power Dissipation

Another important parameter for logic families is power dissipation of the logic gate.

Temperature Range

Another important parameter of logic families, is the operating temperature range, of the logic gate.

Figure of Merit

The product of propagation delay time and the power dissipation is known as **"figure of merit."**

Other Parameters

The other parameters for logic families include suitability for IC fabrication, cost etc.

11.3. SOME LOGIC FAMILIES

Diode Logic Family (DL Family)

In this case, the logic is performed by diodes. AND gates and OR gates can easily be made using DL logic.

Resistor Transistor Logic (RTL)

Here, the logic is performed using resistors and transistors. This is a simple and economical logic family and is best suited for slow-speed industrial and commercial applications. The basic gates are NOR, NAND, and NOT.

Resistor Capacitor Transistor Logic (RCTL)

With this logic, operations are performed by resistors, capacitors, and transistors. The switching speed of RCTL is slower than the switching speed of RTL.

Diode Transistor Logic (DTL)

This is an improved version of the DL family. The fan-out parameter of DTL is higher than that of DL because of the inclusion of an active component, the transistor, in the logic. The basic gates of this family are NAND and NOR.

Direct Coupled Transistor Logic (DCTL)

DCTL is the same as DTL, but the base resistance is removed. The basic gates of this family are NAND and NOR.

High Threshold Logic (HTL)

HTL is the same as DTL except (*i*) a higher supply voltage is given, (*ii*) a zener diode is used instead of an ordinary diode, and (*iii*) large values of resistance are used.

Intregrated Injection Logic (IIL)

This is also called I^2L and its most noble feature is its compact structure and high circuit density.

It uses a PNP transistor and a multi-emitter NPN transistor. The basic gates are AND, NAND, and NOT.

11.4. TRANSISTOR TRANSISTOR LOGIC (TTL)

This logic is also referred to as T^2L. TTL acts fast compared to many other logic families.

Here, multi-emitter transistors that have multiple emitter terminals are used. The advantages of using a multi-emitter structure are listed below:

(*i*) Silicon area is more efficiently used resulting in a higher packing density.

(*ii*) The switching speed is improved.

The standard gate of this family is the NAND gate. The simplest form of a TTL NAND gate is shown in Figure 11.5.

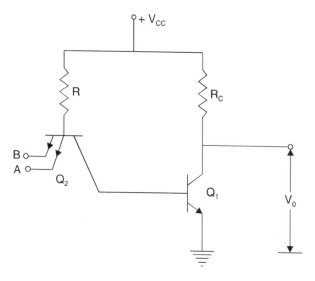

Fig. 11.5. Simple TTL NAND gate.

The standard form of a TTL NAND gate is shown in Figure 11.6.

Standard values:

$$+V_{CC} = +5\ V$$

$$R_1 = 4\ k\Omega$$

$$R_2 = 1\ k\Omega$$

$$R_3 = 130\ \Omega$$
$$R_4 = 1.6\ \text{k}\Omega$$

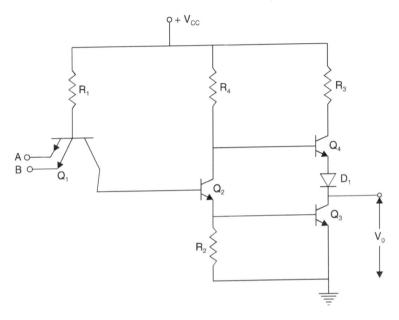

Fig. 11.6. Standard TTL NAND gate.

Let the simplest form of TTL NAND gate structure be used for this explanation.

As an initial step, we will assume that all the inputs are in the logic-0 level. A typical logic-0 level input is 0.3 V. Such an input voltage will forward bias the base junctions and the common base voltage will be 1 V. To forward bias transistor Q_1, the common base voltage must be 1.3 V to 1.4 V. As such, the transistor Q_1 will be in the OFF state and its collector voltage will be V_{CC} volts.

Now we will analyze the operation of the circuit if all inputs are at the logic-1 state, *i.e.*, V_{CC} volts. All emitter-base junctions of transistor Q_2 will be reverse biased. The collector base junction of the multi-emitter transistor and emitter-base junction of Q_1 will be forward biased. Therefore, Q_2 operates in an inverted mode, *i.e.*, with the roles of collector and emitter switched. The common base voltage will be 1.4 V and this will keep transistor Q_1 saturated and the output voltage will be V_{CE} sat.

$$\therefore \qquad V_{iH} = 1.5 \text{ V}$$
$$V_{iL} = 0.1 \text{ V to } 0.3 \text{ V (typically)}$$
$$V_{OH} = 3.5 \text{ V}$$
$$V_{OL} = V_{CE} \text{ sat} \simeq 0.2 \text{ V}$$

The drawback associated with this simple model of TTL NAND gates is that it is not suitable for driving capacitance with low impedance. Loads often arise in high-speed circuits because of the high input impedance. Because of this drawback, this simplest form of TTL NAND gate is seldom used.

The standard form of TTL NAND gate is also called a modified TTL NAND gate. In this circuit, a totem pole or active pull-up stage is added to the simple TTL NAND gate to increase fan-out.

Transfer characteristics of a TTL gate are shown in Figure 11.7.

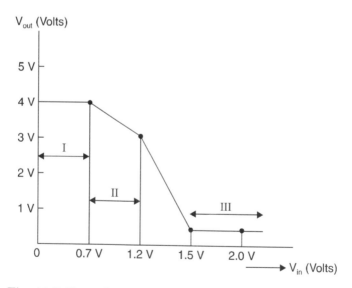

Fig. 11.7. Transfer characteristics of a TTL NAND gate.

Region I is for logic-0 input.

Region II is the transition stage.

Region III is for logic-1 input.

11.5. EMITTER COUPLED LOGIC (ECL)

Also referred to as **Emitter Coupled Transistor Logic (ECTL)** this logic employs an emitter coupled differential amplifier. The basic gates of this family are OR, NOT, and NOR.

This family has the minimum propagation delay because the output is not driven into saturation.

A standard ECL OR/NOR gate is drawn in Figure 11.8.

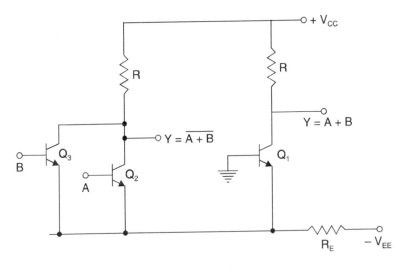

Fig. 11.8. ECL OR/NOR gate.

We will assume that all inputs are equal to logic-0. Therefore, the base of Q_2 and Q_3 will be within the range of -1 to -2 V. Q_2 and Q_3 are in an OFF state.

Q_1 is in an ON state because resistance of Q_1 is less. Therefore, the voltage drop across Q_1 is less and the voltage at the collector of Q_1 = logic-0.

We will consider for the next case that all the inputs are equal to logic-1. Now the base of Q_2 and Q_3 will be within the range of +1 V to + 2 V. Therefore, Q_2 and Q_3 will be in an ON state and Q_1 is in an OFF state. The resistance of Q_1 is high and the voltage drop across Q_1 is high. Therefore, the voltage at the collector of Q_1 = logic-1.

11.6. COMPLEMENTARY METAL OXIDE SEMICONDUCTOR (CMOS) LOGIC

MOS Logic

This logic is similar to the other transistor logic families except it uses metal oxide semiconductor field effect transistors instead of bipolar junction transistors. The advantages offered are easier fabrication process, increased operating speed, and low power comsumption.

Generally, the N-channel MOSFET is used in this family. Hence the logic is named **NMOS logic.** The basic gate used is a NOT gate.

CMOS Logic

Here, N-channel and P-channel MOSFETs are used. This increases the systems complexity and chip area compared to NMOS logic. The great advantage of CMOS logic is that the power consumption in a steady state is almost zero. Power consumption occurs only when there is a switching action from one state to another. The basic gates used are NOT and NAND.

The CMOS NOT gate is drawn and explained in Figure 11.9.

Logic-1 = V_{CC} and Logic-0 $\simeq$ 0 volts.

If V_{in} = 0 volts, then the impedance between the source and drain terminals of the upper P-channel MOSFET is approximately zero. Therefore, no current flows through the upper transistor. Hence, output voltage, V_{out} = V_{CC} = Logic-1.

If V_{in} = V_{CC} volts, then the lower N-channel MOSFET produces a virtual ground at the output. Therefore, output voltage, V_{out} = 0 volts = Logic-0.

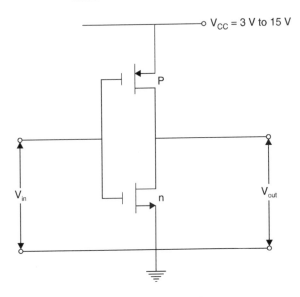

Fig. 11.9. CMOS NOT gate.

A CMOS NAND gate is drawn in Figure 11.10 and explained.

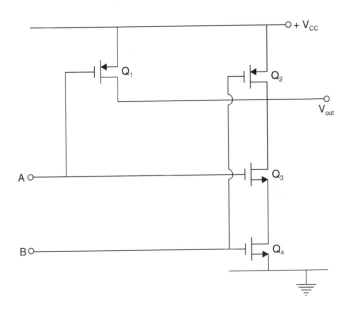

Fig. 11.10. CMOS NAND gate.

Q_1 and Q_2 are *P*-channel MOSFETs

Q_3 and Q_4 are *N*-channel MOSFETs

If $A = 1$, $B = 0$

Q_1 is OFF, Q_3 is ON,

Q_2 is ON, Q_4 is OFF

Since Q_2 is ON, output voltage, $V_{out} = V_{CC}$ = Logic-1.

If $A = 0$, $B = 1$

Q_1 is ON, Q_2 is OFF,

Q_3 is OFF, Q_4 is ON

Since Q_1 is ON, output voltage, $V_{out} = V_{CC}$ = logic-1.

If $A = 0$, $B = 0$

Q_1 is ON, Q_2 is ON,

Q_3 is OFF, Q_3 is OFF

Since Q_1 and Q_2 are ON, output voltage, $V_{out} = V_{CC}$ = Logic-1.

If $A = 1$, $B = 1$

Q_1 is OFF, Q_2 is OFF,

Q_3 is ON, Q_4 is ON

Since Q_1 and Q_2 are OFF and Q_3 and Q_4 are ON, output voltage, $V_{out} = 0$ = Logic-0.

11.7. COMPARISON OF LOGIC FAMILIES

The comparison of the different logic families is shown as a table.

Property	DL	RTL	HTL	DTL	TTL	ECL	MOS	CMOS
1. Basic gates	AND, OR	NOT, AND, NAND	NAND	NAND	NAND	OR, NOR	NAND, NOT	NAND, NOR
2. Fan-out (minimum)	2	5	10	8	10	25	20	50
3. Typical power dissipation in mw per gate	12	12	55	8–12	12–22	40–55	0.2–10	0.01 (Static)
4. Noise immunity	Poor	Medium	Excellent	Good	Very good	Good	Medium	Very Good
5. Typical propagation delay per gate in ns	100	12	90	30	6–12	1–4	300	70
6. Clock rate maximum frequency at which flip flops can operate in MHz	1	8	4	12–30	15–60	60–400	2	5

Property	DL	RTL	HTL	DTL	TTL	ECL	MOS	CMOS
7. Number of functions	Medium	High	Medium	Fairly high	Very high	High	Low	Low
8. Signal generation	Passive	Active	Active	Active	Active	Active	Active	Active
9. Supply voltage	5 V, 10 V	5 V	5 V	5 V	5 V, 10 V	Dual, + 10 V, − 10 V and + 20 V, − 20 V	3–15 V	3–15 V
10. Application	Limited to slow-speed industrial	Slow-speed industrial	Slow-speed industrial	High speed, computer, etc.	Very high speed, Computer, etc.	Very high speed	Slow speed	Slow speed

REFERENCES

1. Mithal, G.K. *Electronic Devices and Circuits*. Khanna, 1994.

2. Choudhury, D. Roy, and Shail Jain. *Linear Integrated Circuits*. New Age, 1989.

3. Muthusubramaniyam, S. Salivahanan and Muralee Dharan. *Basic Electrical and Electronics Engineering*. Tata McGraw Hill, 2000.

4. Gayakwad, Ramant A. *Operational Amplifiers*. PHI, 1994.

5. Coughlin, Driscoll. *Operational Amplifiers and Linear Integrated Circuits*. PHI.

6. Mano, M. Morris. *Digital Design*. PHI, 1994.

7. Sonde, B.S. *Introduction to System Design Using Integrated Circuits*. Wiley Eastern, 1987.

8. Rajaraman, V. Radhakrishnan T. *An Introduction to Digital Computer Design*. PHI, 1991.

9. Bogart, Heodore F. *Introduction to Digital Circuits*. McGraw Hill, 1992.

10. Botkar, K.R. *Intregrated Circuits*. Khanna, 1996.

11. Taub, Herbert, and Donald Schilling. *Digital Integrated Electronics* McGraw Hill, 1977.

12. Gaonkar, Ramesh S. *Microprocessor Architecture, Programming and Applications*. New Age, 1996.

Index